THE ART OF GAME THEORY

HOW TO WIN LIFE'S ULTIMATE PAYOFFS
THROUGH THE CRAFT OF PREDICTION,
INFLUENCE, AND EMPATHETIC STRATEGY

WISDOM UNIVERSITY

CONTENTS

For Our Readers	1
What Reader's Are Saying About Wisdom University	9
Introduction	13
1. Everyday Game Theory - Why Smart People Do Foolish Things	19
2. The Dilemma Of Trust - How To Avoid Jail Time	33
3. The Hidden Chessboard - How Game Theory Shapes Our World	49
4. Making Good Choices - What You Can Learn From The Kindergarten Playground	66
5. Sign Language - When To Fit In And When To Stand Out	80
6. Tactical Retreat - How Stepping Backwards Can Move You Forward	91
7. A Strategic Legacy - Standing On The Shoulders Of Giants	103
8. Practice Makes Permanent - How To Never Forget What You Learn	118
9. Making A Playbook - Why Humans And Robots Are Different	131
Afterword	145
Over 10,000 People Have Already Subscribed. Did You Take Your Chance Yet?	149
The People Behind Wisdom University	151
References	155
Disclaimer	161

Get 100% Discount On All New Books!

Get ALL our upcoming eBooks for FREE
(Yes, you've read that right)
Total Value: $199.80*

You'll get exclusive access to our books before they hit the online shelves and enjoy them for free.

Additionally, you'll receive the following bonuses:

Bonus Nr. 1

Our Bestseller
How To Train Your Thinking
Total Value: $9.99

If you're ready to take maximum control of your finances and career, then keep reading…

Here's just a fraction of what you'll discover inside:
- Why hard work has almost nothing to do with making money, and what the real secret to wealth is
- Why feeling like a failure is a great place to start your success story
- The way to gain world-beating levels of focus, even if you normally struggle to concentrate

★ ★ ★ ★ ★

"This book provides a wealth of information on how to improve your thinking and your life. It is difficult to summarize the information provided. When I tried, I found I was just listing the information provided on the contents page. To obtain the value provided in the book, you must not only read and understand the provided information, you must apply it to your life."

NealWC - Reviewed in the United States on July 16, 2023

"This is an inspirational read, a bit too brainy for me as I enjoy more fluid & inspirational reads. However, the author lays out the power of thought in a systematic way!"

Esther Dan - Reviewed in the United States on July 13, 2023

"This book offers clear and concise methods on how to think. I like that it provides helpful methods and examples about the task of thinking. An insightful read for sharpening your mind."

Demetrius - Reviewed in the United States on July 16, 2023

"Exactly as the title says, actionable steps to guide your thinking! Clear and concise."

Deirdre Hagar Virgillo - Reviewed in the United States on July 18, 2023

"This is a book that you will reference for many years to come. Very helpful and a brain changer in you everyday life, both personally and professionally. Enjoy!"

Skelly - Reviewed in the United States on July 6, 2023

Bonus Nr. 2
Our Bestseller
The Art Of Game Theory
Total Value: $9.99

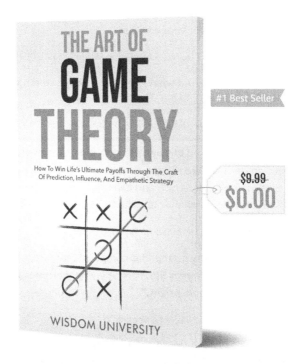

If Life is a game, what are the rules? And more importantly... Where are they written?

Here's just a fraction of what you'll discover inside:
- When does it pay to be a selfish player... and why you may need to go inside a prisoner's mind to find out
- How to recognize which game you're playing and turn the tables on your opponent... even if they appear to have the upper hand
- Why some games aren't worth playing and what you should do instead

"Thanks Wisdom University! This book offers simple strategies one can use to achieve things in your personal life. Anyone of average intelligence can read, understand and be in a position to enact the suggestions contained within."

David L. Jones - Reviewed in the United States on November 12, 2023

"Haven't finished it yet, but what I've gone through so far is just incredible! Another great job from this publisher!"

W. S. Jones - Reviewed in the United States on October 12, 2023

"A great book to help you through difficult and complex problems. It gets you to think differently about what you are dealing with. Highly recommend to both new and experienced problem solvers. You with think differently after reading this book."

Thom - Reviewed in the United States on October 18, 2023

"I like this book and how it simplifies complex ideas into something to use in everyday life. I am applying the concept and gaining a lot of clarity and insight."

Ola - Reviewed in the United States on October 18, 2023

"The book is an excellent introduction to game theory. The writing is clear, and the analysis is first-rate. Concrete, real-world examples of theory are presented, and both the ways in which game theory effectively models what actually happens in life is cogently evaluated. I also appreciate the attention paid to the ethical dimensions of applying game theory in many situations."

Amazon Customer - Reviewed in the United States on October 8, 2023

Bonus Nr. 3 & 4

Thinking Sheets
Break Your Thinking Patterns
&
Flex Your Wisdom Muscle
Total Value Each: $4.99

A glimpse into what you'll discover inside:
- How to expose the sneaky flaws in your thinking and what it takes to fix them (the included solutions are dead-simple)
- Dozens of foolproof strategies to make sound and regret-free decisions leading you to a life of certainty and fulfillment
- How to elevate your rationality to extraordinary levels (this will put you on a level with Bill Gates, Elon Musk and Warren Buffett)
- Hidden gems of wisdom to guide your thoughts and actions (gathered from the smartest minds of all time)

Here's everything you get:

- ✓ How To Train Your Thinking eBook — ($9.99 Value)
- ✓ The Art Of Game Theory eBook — ($9.99 Value)
- ✓ Break Your Thinking Patterns Sheet — ($4.99 Value)
- ✓ Flex Your Wisdom Muscle Sheet — ($4.99 Value)
- ✓ All our upcoming eBooks — ($199.80* Value)

Total Value: $229.76

Go to wisdom-university.net for the offer!

(Or simply scan the code with your camera)

*If you download 20 of our books for free, this would equal a value of 199.80$

WHAT READER'S ARE SAYING ABOUT WISDOM UNIVERSITY

"Wisdom University embodies an innovative and progressive educational approach, expertly merging deep academic insights with contemporary learning techniques. Their books are not only insightful and captivating but also stand out for their emphasis on practical application, making them a valuable resource for both academic learning and real-world personal development."

—*Bryan Kornele, 55 years old, Software Engineer from the United States*

"I associate Wisdom University with critical thinking and knowledge improvement. It is helpful for critical thinkers and all those who are interested in improving their knowledge."

—Elliot Wilson, MBA and Doctor of Business Administration (DBA), Chief Growth Officer

"I wanted to read some books about thinking and learning which have some depth. I can say "Wisdom University" is one of the most valuable and genuine brands I have ever seen. Their books are top-notch at kindle. I have read their books on learning, thinking, etc. & they are excellent. I would especially recommend their latest book "Think Like Da Vinci" to those who want to have brilliant & clear thinking."

—*Sahil Zen, 20 years old from India, BSc student of Physics*

"I would recommend these books to my grandson."

—Daniel, Florida (USA), 69, Bachelor Degree, retired

"I have been reading books from Wisdom University for a while now and have been impressed with the CONDENSED AND VALUABLE INFORMATION they contain. Reading these books allows me to LEARN INFORMATION QUICKLY AND EASILY, so I can put the knowledge to practice right away to improve myself and my life. I recommend it for busy people who don't have a LOT of time to read, but want to learn: Wisdom University gives you the opportunity to easily and quickly learn a lot of useful, practical information, which helps you have a better, more productive, successful, and happier life. It takes the information and wisdom of many books and distills and organizes the most useful and helpful information down into a smaller book, so you spend more time applying helpful information, rather than reading volumes of repetition and un-needed filler text.

—*Dawn Campo, Degree in Human psychology and Business, Office administrator from Utah*

"Wisdom University's works provide a synthesis of different books giving a very good summary and resource of self-help topics. I have recommended them to someone who wanted to learn about a topic and in the least amount of time."

—Travvis Mahrer, BA in Philosphy, English Teacher in a foreign country

"I have most of the ebooks & audiobooks that Wisdom University has created. I prefer audiobooks as found on Audible. The people comprising Wisdom University do an excellent job of providing quality personal development materials. They offer value for everyone interested in self-improvement."

—*Neal Cheney, double major in Computer-Science & Mathematics, retired 25yrs USN (Nuclear Submarines) and retired Computer Programmer*

"WU's emails discuss interesting topics. They have good offers. I can recommend the books to my My friends and relatives."

—Wilbur Dudley, Louisiana (USA), 77, BS in Business Administration and DBA, retired

INTRODUCTION

In September 2000, Reed Hastings and Marc Randolph were laughed out of the most important meeting of their lives.

Three years earlier, Reed and Marc had founded a company that let customers order DVDs for rental via the Internet. They called their firm Netflix. However, with the dot-com bubble collapsing, Netflix was in serious financial trouble.

Reed and Marc had been trying to get a meeting with Blockbuster executives for months. Given the environment, a sale to a large competitor looked like one of the few ways for the firm to survive. They were finally granted a short audience at Blockbuster's corporate headquarters in Dallas.

At the meeting, Reed and Marc proposed selling Netflix to Blockbuster for $50 million, believing that Blockbuster could effectively leverage Netflix's business model and brand. By all accounts, the Blockbuster executives thought the proposal was a joke. They categorically rejected the offer and sent Marc and Reed packing.

The rest is history. Netflix managed to survive and, by 2010, was valued at more than $3 billion. That same year, Blockbuster went bankrupt, despite formerly being one of the leading players in the DVD-rental space.[1] In a short time, Netflix went from being treated as a joke to becoming one of the most influential companies in the world.

Strategy—one of the most important elements of all, made up for what Netflix lacked in respect or resources. By identifying the potential of the Internet, cutting down on annoying fees, and employing a capital-light business model, Netflix overthrew the king of the movie-rental industry.

The Importance Of Strategy

The Netflix story contains a vital life lesson that anyone in the business world should appreciate. While the biggest, strongest, and wealthiest competitors have a leg up on the rest, effective use of strategy can turn even the smallest opponent into a serious threat.

In your own life, you may be facing similar competitive struggles. Perhaps your peers are making consistent advancements, while you're stuck in place. Or perhaps you're moving forward, but not as quickly as you'd like. If you're pursuing business or educational ambitions, you may be struggling to balance your personal responsibilities, perhaps spending less time with your family and friends than you'd like.

If this sounds familiar, you're facing the same difficulties as many other ambitious people. Even though some doubt it, it is absolutely possible to lead a balanced, fulfilled life, where you achieve your career ambitions without neglecting the things that make life worth living. But to get there, you'll need to understand how to leverage strategic thinking to make the most effective use of your time.

Through this book, we'll show you how the field of game theory can help you do just that.

What You'll Learn

Although game theory can be an abstract field, this is not a theoretical work. Our approach is to synthesize the key findings of the field into simple lessons that you can put into practice immediately.

Game theory is an expansive field whose development has been supported by some of the brightest minds in history. First, we'll guide you through the way in which game theory understands the world. Breaking down situations into clear structures can help cut through the opacity and noise that dominates the real world.

Then, we'll see how various strategies, inspired by decades of research, can help you navigate these structures. The recent explosion in popularity surrounding the use of laboratory experiments to test economic theories will be a great benefit to us. While previous works on game theory might have included strategies that sound convincing on paper, we get to analyze which strategies actually work in the real world.

Now, thinking about how you think can be difficult, which is why so many people avoid it. It's easy to be complacent about our own approach to things, especially when we don't know about other possible approaches. But complacency is an easy way to keep getting the results you are, while your competitors streak ahead.

Others might worry that it's not the right time in their life to begin this strategic journey. One lesson from game theory, though, is that the world around us is always evolving. There will never be a perfect time to start working on your problems, which means your best bet is to start today.

Are you wondering whether strategic thinking can make a genuine difference? If you're looking for a real case study, look no further than the author of this book.

My Game Theory Journey

Learning to think strategically made a profound impact on my life.

Having graduated from the University of Virginia with a degree in Economics, I already had some exposure to game theory. However, I did not truly recognize its importance until I began working in the real world.

When I began my first job in finance, I spent a significant amount of time ensuring that my models and calculations were the most advanced in the office. In school, I had learned cutting-edge financial theory, and was eager to put it to work.

Over time, however, I recognized that simply optimizing this one aspect of my job was not getting me results. Clients and colleagues were rarely making decisions based on which financial model utilized the latest research.

I started to think a bit more strategically about the way I operated. This included spending my time understanding people better, rather than on math, and learning how to integrate my goals with other peoples' rather than viewing them as strictly separate.

This approach paid off for me, and I began to garner responsibility at a quicker pace. This culminated in my being named Chief Strategist at the firm, and being able to incorporate game theory and strategic thinking for the success of the business as a whole.

Now, having stepped away from the financial world, I continue to leverage strategy as a freelance writer. Without learning how to navigate the competitive world of finance, I could not have made the leap to pursue a new career path.

If strategic thinking can work for me, it can certainly work for you. Are you ready to learn how to use game theory to build the life you want? Let's get started.

Brian Flaherty

1
EVERYDAY GAME THEORY - WHY SMART PEOPLE DO FOOLISH THINGS

Mark stumbled out of the bar, his girlfriend following close behind him.

Shortly after, Sean emerged onto the street, surrounded by several friends. As his friends muttered words of encouragement, Sean threw several warm-up punches into the cool night air.

In their semi-drunken haze, neither Mark nor Sean could recall the exact string of events that led to this confrontation. Drinks had been spilled, and insults exchanged. Now the two were convinced that fighting was the only way to defend their respective honor.

Mark and Sean are both intelligent men who know full well the brutal and dangerous nature of street fights. Even the winner of the fight could leave with a broken hand or a busted tooth. While neither man wants to fight, they also can't afford to back down and look like a coward.

Both men have been hoping the other will be the one to apologize and ask for mercy before the fight starts. Even for the loser, backing down is optimal since a wounded ego is preferred to a wounded body. Yet as the fight draws near, both Mark and Sean refuse to yield.

The Logic Of Mutual Destruction

Whether or not you find yourself frequently engaging in bar fights, you've probably encountered situations in which otherwise intelligent people find themselves on a collision course toward mutual destruction. Consider two arguing lovers, each unwilling to be the one to apologize, until an otherwise happy relationship breaks apart. On a grander scale, consider two rival superpowers, each refusing to back down from conflict, even at the risk of nuclear Armageddon.

Surprisingly, despite their superficial differences, all three of these scenarios represent very similar games. The fighters, the lovers, and the superpowers are all playing versions of 'chicken.' In chicken, named after the game where two drivers head straight toward each other on a collision course, both participants can avoid a disastrous outcome as long as one of them yields. But since backing down can bring on personal shame (like being labeled a coward or a pushover), both participants gamble that the other will yield first.

Stripping away the narrative and focusing on payoffs won't win us any points if we're telling a story at a party, but it should help us understand the mathematical approach to decision-making that underlies game theory. In this chapter, we'll start thinking like a game theorist by analyzing real-life situations or stories in terms of their competitive structure.

Thankfully, not all games are as depressing as a lover's quarrel or nuclear tensions. We'll start by differentiating between non-zero-sum games, where 'win-win' situations might exist, and zero-sum games, where the winner benefits at the expense of the loser.

Zero-Sum And Non-Zero-Sum Games

When people first hear about game theory, their mind typically jumps to an analysis of either sporting competitions or board games, depending on their temperament. While the field is much more expansive than merely studying literal games, both sports and board games provide excellent examples of 'zero-sum' games.

In zero-sum games, typically played by two participants, one side can only win at the expense of the other. The vast majority of board, party, and card games are zero-sum affairs. Consider chess. Players can either win, lose, or draw, but benefit to one side can only accrue due to a loss to the other. Even in Risk, a strategy board game of territorial conquest, alliances cannot last. Eventually, only one player is left standing.

Sports games tend to result in similar dynamics, with each team trying to win, necessarily meaning the other team loses. Soccer, cricket, baseball, football, and hockey are all generally zero-sum, for example.

An interesting exception is when the structure of a tournament makes two teams desire the same final result. For instance, in the 1982 FIFA World Cup match between Germany and Austria, both teams knew that they could advance to the next round safely so long as Germany won the match by one or two goals. After Germany scored one goal at the beginning of the game, both teams engaged in a half-hearted kickaround until time expired. Austria may have lost the match, but both teams benefited by easily advancing to the next round (at the expense of Algeria, who lodged an unsuccessful complaint).[1]

The Germany-Austria game, then, is a rare example of a non-zero-sum sports game. Non-zero-sum games occur when one player winning does not necessarily mean the other player loses (and vice versa – one player losing does not mean the other player wins).

Many games in the business world are non-zero-sum, or at least contain non-zero-sum elements. At first glance, wage negotiations between a union and a firm may appear strictly zero-sum, with pay raises to the union resulting in lower profits for the firm. While the game is zero-sum so long as a solution is eventually found, if no wage agreement occurs, then workers will be out of a job, and the firm will be out of business. For this reason, while the 2023 Hollywood writers' strike is contentious, neither party has an interest in seeing the opposing side lose unconditionally.

Interestingly, while participants in the chicken game described earlier are opponents, the game is not actually zero-sum. If both participants refuse to yield, then both will lose, but neither will win. The best way to differentiate a zero-sum game from a non-zero-sum game is to ask the following question: if one participant loses, is it *always* the case that the other participant wins?

Cooperative And Non-Cooperative Games

In the 2010 film *Inception*, the powerful businessman Saito promises the protagonist Cobb that if Cobb completes one last job, Saito will arrange for Cobb's criminal record to be expunged. A major plot point revolves around Saito's reliability – if the businessman does not honor the arrangement, Cobb faces a life sentence. In this situation, Cobb and Saito are playing a non-cooperative game; Cobb has no way of enforcing Saito's promise, and Saito offers no commitment mechanism.

This example is a non-cooperative game, as opposed to a cooperative one. Despite what the names imply, the difference between cooperative and non-cooperative games has little to do with how amenable to collaboration participants are. Nor does it have to do with how aligned their interests are. Rather, cooperative games feature *binding agreements*, while non-cooperative games feature only *non-binding agreements*.

In modern research, the study of non-cooperative games is more popular than cooperative games. This is because it's far more realistic to assume that any agreement is non-binding and therefore needs an enforcement mechanism. It's exceedingly rare for cooperation to be simply based on trust, especially in politics and business. Usually, a penalty must be applied to any competitor who breaks the agreement.

What makes game theory special is the way that our optimal strategy depends on other people's choices. Were it not for this consideration, game theory could be replaced by a simple expected value calculation. Therefore, it's unsurprising that the existence of true promises is important enough to be categorized in its own class of games. What matters, though, is not just what others do, but whether we learn about their choices in time to factor into our own decision-making.

Sequential And Simultaneous Games

Sequential and simultaneous games serve as the next broad categorizations that games fall into. In sequential games, participants make their strategic choices one by one. In simultaneous games, as you'd expect, all participants make their choices at the same time.

The result is that participants in sequential games operate with clear information about the choices of other players. Some games have to operate sequentially in order to function; imagine trying to play simultaneous chess.

Other games can be played sequentially or simultaneously, and therefore have to be analyzed as distinct scenarios.

For instance, suppose you've been invited to a real estate auction to bid on your dream home. The bidding is structured like a standard open-bid auction, with bids increasing until no higher offer is put forth. This is clearly a sequential game since you get to observe and react to your competitor's offers. As long as you value the home more than anyone else (and can afford to pay), the dream home will invariably be yours in the sequential game.

Suppose, however, the auctioneers structure the proceedings as a sealed-bid auction. In this instance, all bidders write their offers in a sealed envelope. The envelopes are opened, and the highest bidder gets the house. Since you don't get to see the choices of your competitors, this is a simultaneous game.

Now, you might be torn over what to bid in the sealed auction. Bid too high, and you risk paying far more than you would have in the open-bid auction. Bid too low, and you risk not getting your dream home. The amount you bid will depend on what you expect your competitors to bid, an unobservable variable in this new simultaneous game.

Simultaneous sealed-bid auctions, in addition to requiring a different analysis than an open-bid auction, are also significantly less entertaining than their sequential counterparts. Imagine how dull the auction scene in the Hitchcock classic *North by Northwest* would have been if

Roger Thornhill made absurd bids in private, rather than in public.

Clearly, introducing the concept of time into games can alter our analysis, since the second mover in a sequential game can learn about their opponent's strategy before making their own choice. In the next section, we'll extend this idea to understand how playing a game multiple times can lead to different results than playing the same game just once.

One-Shot And Repeated Games

In a classic episode of the animated sitcom *The Simpsons*, a smooth-talking salesman convinces the people of Springfield to spend millions of dollars on a new monorail system. Unbeknownst to the townspeople, however, the monorail is faulty, and the salesman absconds with the cash skimmed from the project.

Now, suppose the salesman showed up to Springfield again the next year, promising a functioning monorail system in exchange for a fresh investment. If the citizens had any sense, they would kick the con man out of town, recognizing him as a scam artist. By learning from their past mistakes, the people of Springfield can update their choices in the future.

This example highlights the important distinction between playing a game once and playing it repeatedly. The idea that tricking people twice is far more difficult than tricking them once (and that people are expected to learn from past experience) is captured in the classic saying, "Fool me once, shame on you. Fool me twice, shame on me."

In game theory, games that are played once are known as one-shot, while games played multiple times are known as repeated. Additional distinctions can be made for games repeated a finite number of times versus an infinite number of times. Moreover, whether or not participants know the total number of games to be played can factor into their decision-making.

For example, we learn about other people's traits through repeated interaction, and factor in our own status when we know reputation is on the line. As social creatures, reputation is one of the oldest games that human beings play. This is why you trust an old friend more than a new acquaintance. Not only have you played many different games with your friend over the years, allowing you to learn about their honesty and loyalty, but you expect to interact with them again in the future, making reputation a factor in their decision-making.

In general, our capacity to learn from past experience to inform our future decisions creates distinct dynamics for one-shot and repeated versions of the same game.

Game Theory As Art: The Map Is Not The Territory

While the foregoing discussion may have presented games as clearly delineated, and game theory as providing precise answers, reality is a bit messier. Game theory is more of an art than science, precisely because real-world environments will not match the well-defined mathematical setups or laboratory experiments created in a research environment.

Any abstract game can be exactly defined in an academic paper, complete with known payoffs and rational participants capable of calculating expected values on the fly. In the real world, of course, ambiguity reigns. Payoffs and probabilities can be little more than educated guesses. Sufficient life experience will disabuse any reader of the notion that people are always rational.

Ultimately, though, game theory can still help us navigate the real world. Just because an academic concept does not precisely correspond to a real-world situation does not mean the concept isn't useful. For instance, the rule that the shortest route between two points is a straight line technically only holds for flat planes, but the rule works well enough when walking around the surface of a large oblate spheroid that we use it in our everyday lives.

Studying game theory will provide us with general rules for navigating strategic decision-making in the real world. But the field is an art, not a science, because knowing when to break the rules is just as important as knowing when to follow them.

Action Steps

To engage with the concepts presented in this chapter, and to prepare yourself to start thinking like a game theorist, try and answer the following questions. For those who want to get the most out of this book, writing out explanatory reasoning for each answer is a useful practice.

1. Each year, a mob boss negotiates a bribe with a crooked police commissioner. The two discuss the terms of the bribe over a steak dinner, and negotiations can last for several hours. Using the four distinctions proposed in this chapter, what kind of game is being played?

2. You are the president of a country serving your last term allowable by law. During this final term, you want to aggressively enact your legislative vision, especially for policies unpopular with the opposition. Using ideas discussed in this chapter, why might you be more willing to pursue legislation aggressively in your final term, rather than earlier ones?

3. You've set up a joint international business venture in a developing country with an unreliable court system. Your local partner signs a contract that they claim binds them to carry out their commitments, or face severe legal punishment. Which aspects of this game are cooperative, and which are non-cooperative?

4. What are the possible limitations of game theory when applying the field to the real world?

If You've Got A Bazooka…

When we last left Mike and Sean, they were continuing their dangerous game of chicken, each hoping the other would back down before the fight started. Now that we've reviewed some different game structures, and understand game theory as the art of putting yourself in your opponents' shoes, what advice might we have for the two men?

Consider that the fight would not occur so long as one of the men had a binding commitment to fight. If Mike were convinced that Sean would *never back down*, Mike would recognize his optimal strategy as yielding, even at the cost of a bruised ego. In the real world, of course, no such binding commitments exist, as we discussed in the cooperative game section of this chapter.

Still, Sean might be able to make a sufficiently convincing commitment that Mike evaluates the probability of Sean backing down as nearly zero. For instance, suppose Sean came from a culture with an extreme disdain for cowardice or retreat. Alternatively, Sean might be wearing a t-shirt from the local boxing gym, indicating he fights on a regular basis. Indicating a credible capacity for violence can sometimes be the best way to avoid real instances of violence.

Now that we understand the basics of game theory, we'll start to dive deeper into some of the most studied games in the field. As we'll find out, though, while knowing how to play certain games is important, knowing which games to avoid playing can be just as vital.

Key Takeaways

- Zero-sum games exist when the winner always gains at the loser's expense. This is common in sports or board games. Non-zero-sum games occur when both parties can be winners or when both parties can be losers. These games are more common in business or politics.
- Cooperative games exist when participants can make binding commitments to certain actions. That is, they can honestly tell their opponents they will play a certain strategy with 100% certainty. Since truly binding commitments are exceedingly rare in the real world, and since cooperative games can always be modeled as non-cooperative ones, the study of non-cooperative games dominates in game theory.
- Sequential games occur when participants choose their moves one after the other. Simultaneous games, on the other hand, involve participants making moves at the same time. In sequential games, the second mover has information about the first mover's strategy, which is not the case in simultaneous games.
- In one-shot games, a game is played just once. In repeated games, a certain game is repeated with the same participants over the course of multiple rounds. Sometimes, repeated games can even be infinite.

- Game theory is more of an art than a science since rational agents competing amongst known payoffs and probabilities is an approximation of the real world.

2
THE DILEMMA OF TRUST - HOW TO AVOID JAIL TIME

Why is it that world leaders, shortly after making a long speech about the importance of addressing climate change, will ride in a gas-guzzling motorcade to the airport, where their private plane awaits to take them on an intercontinental flight?

Perhaps the behavior can be justified on the grounds of personal security or the high value of an influential decision-maker's time. Still, the fact remains, that while such people might recognize the importance of fighting climate change, they are unwilling to sacrifice certain personal comforts, even if those comforts lead to unnecessary carbon emissions.

Although the global impacts of climate change might be very high, the individual cost each person bears as a result of a marginal increase in their carbon emissions is staggeringly low. When weighing this small cost against the immediate benefits resulting from time or money saved, the choice is easy. Yet people who make such choices can still be quick to lecture others about climate change, since they recognize that if *everyone* made the same choice as them, the emissions impact could no longer be ignored.

At a personal level, you might have a close friend who is an ardent environmentalist, and yet drives a gas-powered vehicle. To defend themselves, they might cite 'range anxiety,' the high cost of electric vehicles, or some similar discomfort. Although this behavior might seem contradictory, there is a hidden logic driving these decisions that leads to one of the most influential and well-studied scenarios in game theory.

The Prisoner's Dilemma

Whenever a situation is structured such that the optimal choice for an individual is not the optimal choice for *all* individuals to make, some version of the prisoner's dilemma is likely being played.

The prisoner's dilemma is one of the most famous and controversial games in all of game theory. The game takes its name from a motivating story in which two accomplices are arrested for a crime they committed. While the police lack sufficient concrete evidence for a serious conviction, they give each criminal the same offer: testify against your accomplice, and we'll let you walk away scot-free.

Consider the situation. If both criminals manage to stay silent, they may each serve a year in jail before being released. If one criminal testifies against the other, the snitch will walk away with no jail time, while the scapegoat will serve ten years in jail, bearing sole responsibility for the crime. If both criminals testify against the other, they will each serve five years in jail, sharing the responsibility.

Clearly, both accomplices staying silent will lead to the best outcome overall, with a total jail term of just two years altogether. However, each criminal has a dominant incentive to testify. If they are convinced their partner will stay silent, then they should testify to get no jail time. If they are convinced their partner will testify, then they should testify to reduce their jail sentence from ten years to five years. Inevitably, this will lead to both criminals testifying, and failing to achieve the best outcome overall.

In this chapter, we'll explore this example and other variants of the prisoner's dilemma. Despite being formally developed in the buildup to the Cold War, recommendations for navigating similar structures can be found in ancient Roman and Greek sources. The prisoner's dilemma is popular to study precisely because it has been relevant to human decision-making for so long and will continue to be relevant far into the future.

History Of The Dilemma, The Dilemma In History

The prisoner's dilemma was originally formulated in 1950 by Merrill Flood and Melvin Dresher, colleagues at the RAND Corporation, a think tank created to advise American military and security policy in the post-World War II period. RAND was keenly interested in game theory, which was growing into a formalized field of study after the publication of the seminal *Theory of Games and Economic Behavior* by Oskar Morgenstern and John von Neumann in 1944.

While Flood and Dresher's original experiment contained the essential elements that make the prisoner's dilemma so interesting, the presentation was somewhat convoluted and not widely read. It was not until Albert Tucker, a RAND consultant and Princeton mathematician, translated the experiment into a motivating scenario involving criminal confessions that the prisoner's dilemma earned its name.[1] As they say, marketing is everything – the story took on a life of its own and became an influential narrative to describe social dilemmas.

Although the importance of precisely defining the prisoner's dilemma should not be discounted, versions of the dilemma have motivated studies of ethics throughout history. The importance of cooperation for the social good is a primary concern in both religion and philosophy.

Thales of Miletus, a pre-Socratic philosopher who, among other achievements, made one of the first financial derivative trades in recorded history, is attributed with the saying "avoid doing what you would blame others for doing."[2] This sentiment, which mirrors the Golden Rule espoused by Jesus Christ during the Sermon on the Mount, is precisely the type of moral rule meant to promote cooperation in the prisoner's dilemma.

Many interpretations of Immanuel Kant's categorical imperative, which states that you should act in such a way that you would want other people to act the same way,[3] also dictate cooperating in the prisoner's dilemma. While none of these great moral authorities are likely to be specifically concerned with the plight of guilty criminals trying to minimize their jail time, the struggle between cooperation and selfishness throughout history underlines the importance of the prisoner's dilemma.

The Depositor's Dilemma

Despite the name, most people are far more likely to experience a prisoner's dilemma in their business dealings than they are in a police station. Bank runs serve as a classic economic prisoner's dilemma, where mutual cooperation is beneficial, but unlikely.

During the spring of 2023, distress in the American financial system led to a number of large bank failures.[4] While some of these failures were triggered by legitimate business mistakes, others were the result of rapid depositor withdrawals that mirrored the selfish choice in the prisoner's dilemma.

Suppose you were a depositor at a dubious American bank in early 2023, and you were trying to decide whether or not to pull your money out of the bank. If you choose to move your money, and the bank does not collapse, your only loss is the small transition cost you had to pay to switch banks. If you choose to move your money, and the bank does collapse, you will be grateful that you switched banks at the right time. Given this calculation, you are very likely to pull your money out, at least for peace of mind.

The problem, of course, is that if every depositor makes this same calculation, they will all pull their money out in a very short time frame. As anyone who has seen the Christmas classic *It's a Wonderful Life* knows, this simultaneous withdrawal is overwhelmingly likely to lead to the bank's immediate collapse. If all depositors would leave their funds in the bank, the firm could probably muddle through and survive. If all the depositors withdraw, however, the bank will collapse, with potential knock-on effects throughout the financial system. Cooperation might be best for everyone, but it is unlikely to occur.

Note that this example should also highlight that while ethical judgments are related to the prisoner's dilemma, selfish behavior is not necessarily unethical. Technically, depositors who withdraw funds during a bank run are making a selfish decision, but few would cast moral aspersions on them for doing so. The link between selfishness and self-preservation motivates the prisoner's dilemma, after all.

The Dangers Of Dependence

The prisoner's dilemma characterizes one of the elements of life that motivates the study of game theory as a whole; that is, the way our individual outcomes depend on how other people act. While this is particularly important to business considerations, with firms and consumers interacting in the marketplace, it is perhaps more important for political considerations.

Particularly, the prisoner's dilemma can help us determine situations in which society can benefit by entrusting mandatory cooperative mechanisms to the government. While different schools of thought disagree on the extent to which these mechanisms interfere with individual liberty, the prisoner's dilemma is a strong argument that selfish behavior will necessarily preclude socially optimal outcomes.

For instance, consider the climate change example referenced earlier. There is a strong incentive to act selfishly, especially if you believe that everyone else will make the necessary sacrifices to stop climate change. From a personal perspective, refusing to contribute to the social good and reaping the benefits of other people's sacrifices is the best outcome there is. Such a person is known by academics as a 'free rider,' named for people who hop turnstiles on public transport systems to literally ride for free.

A fascinating study about dealing with free riders was undertaken by researchers from Brown University and Tufts University.[5] The research studied 'voluntary contribution mechanisms,' or VCMs. With a VCM, people can pay a voluntary amount toward some public good, which can be enjoyed by all the participants. In the real world, a public good might be a park open to everyone, national defense, or the ability to live on a habitable planet. In the experiment, the researchers simulated a public good by scaling up the total public contributions by about 60%, and then distributing the amount equally to all participants.

The participants were given an initial endowment of money and told that each round, they could contribute some money to the public good, while keeping the rest for themselves. The twist, though, was that participants were also allowed to vote on punishment mechanisms. If the group implemented punishment of low public good contributors, for instance, free riders would have their own earnings reduced.

To begin with, groups were hesitant to punish anyone, with the majority of groups prohibiting punishment altogether. Over time, however, more and more groups implemented punishment mechanisms for free riders, until punishment outweighed non-punishment by nearly four times.

As the researchers note, people may be unwilling to punish others from the outset, but quickly get frustrated with the burden imposed by free riders. Therefore, they eventually punish low contributors to get them to pay their fair share.

The punishment of free riders essentially results in mandatory contributions to the public good, in other words, taxation. This research, therefore, provides an interesting starting point for understanding taxation (and the very existence of government) as a necessary response to the lack of natural cooperation in the prisoner's dilemma.

The Intersection Of Math And Economics

The prisoner's dilemma, perhaps more so than any other game, requires understanding the mathematical decision-making framework that serves as the foundation of game theory. While we aim to limit the extent of mathematical intrusion in this book, the prisoner's dilemma provides a useful scenario through which to understand key concepts in game theory.

Game theorists like to analyze scenarios through what is called a *payoff structure*. Importantly, in the academic literature, these structures define the games themselves, with narratives bolted to the math. This is in contrast to our narrative-first approach, which, while easier to understand, is more difficult to calculate around.

In the prisoner's dilemma previously described in this chapter, the payoff structure looks like this.

Prisoner's dilemma

(A, B)	Cooperate	Testify
Cooperate	(-1, -1)	(-10, 0)
Testify	(0, -10)	(-5, -5)

Prisoner A's choices are on the left, and prisoner B's choices are on the top. In other words, if A chooses to cooperate (stay silent) but B chooses to testify, then A will have a payoff of -10 (representing 10 years in prison) while B will have a payoff of zero (representing no years in prison).

This table helps us understand why testifying is a 'dominant strategy' (a strategy you will play regardless of how your opponent plays). Suppose prisoner A knows that prisoner B will cooperate. Then the only part of the table we are concerned with is the first column. Clearly, a payoff of zero is better than a payoff of -1, so A will choose to testify. Similarly, suppose prisoner A knows that prisoner B will testify. Then we are only concerned with the second column. Since -5 is greater than -10, A will still choose to testify.

Since both prisoners will choose to testify, the bottom right square of (-5, -5) resulting from (Testify, Testify) is known as a 'Nash equilibrium.' This concept, named after famous game theorist John Nash, refers to a situation in which neither player has an incentive to change their strategy, given what the other player has done. Clearly, (Testify, Testify) is a Nash equilibrium in the prisoner's dilemma, since neither player will cooperate knowing that their opponent defected.

It is important to distinguish Nash equilibriums from 'socially optimal' states. Social optimums occur when total welfare is maximized. In the prisoner's dilemma, that is (Cooperate, Cooperate), leading to a total welfare of -2. This is higher than any other state on the table, but, as readers well know by now, will never be reached.

On Playing Unwinnable Games

Given the discussion in this chapter, the outlook looks bleak for human cooperation. If being selfish is a rational strategy in the prisoner's dilemma, what hope is there for working together?

First, it should be noted that the dilemma can easily be modified to induce cooperation by adjusting the payoffs. If there are large enough benefits to cooperation, people will cooperate; but the game they're playing will no longer be the prisoner's dilemma.

Second, defecting is a *rational* choice, but that does not mean it is the realistic choice. The past two decades of groundbreaking work in behavioral economics have proved that people are far from the rational calculation machines that game theory presumes. In fact, some laboratory experiments testing cooperation in games that mirror the prisoner's dilemma found cooperation rates of between 40% and 50%.[6] Clearly, not everybody is the type of cold-blooded, shrewd calculator posited by game theorists.

Third and finally, cooperation may not be a rational strategy in the one-shot version of the prisoner's dilemma, but the story is more complicated for repeated versions. In a repeated prisoner's dilemma, the same participants play over and over, and can therefore learn from their past interactions. One of the best strategies in this form of the game is the remarkably simple 'tit for tat' strategy. Tit for tat advises cooperating initially and defecting only if your opponent defected on the previous move – this serves as a form of short-term punishment. If your opponent goes back to cooperation, then tit for tat advises that you begin cooperating again.

While it is true that there is no way for hyper-rational participants to achieve cooperation in the tightly-defined, one-shot version of the game popular in the academic literature, the real world is far more complicated. With that being said, a great way to avoid jail time is to avoid playing the prisoner's dilemma at all.

Action Steps

To understand the prisoner's dilemma and some of the basic analytical tools of game theory, try and answer the following questions.

1. Suppose you are the CEO of a small bank on the verge of experiencing a bank run. How might you adjust the incentives for depositors to get them to stay? You might benefit from creating an initial payoff structure, then adjusting it until withdrawing is no longer a dominant strategy.

2. Review the following payoff structure for the game of chicken. Remember that a dominant strategy is the optimal strategy for a player regardless of what their opponent does. Does this game have a dominant strategy?

Chicken

(A, B)	Peace	Conflict
Peace	(0, 0)	(-1, +1)
Conflict	(+1, -1)	(-10, -10)

3. Remain with the chicken payoff structure. Recall that a Nash equilibrium is a state in which neither player has an incentive to change strategy, given what the other player has done. Are any of the states in the chicken game a Nash equilibrium? (Hint: There can be more than one.)

4. The prisoner's dilemma highlights the difficulty of cooperation due to people's self-interest. Why, then, does cooperation occur every day in businesses and universities? How are these scenarios different from the pure prisoner's dilemma? If you are managing a group that is not cooperating well, how can the prisoner's

dilemma inform potential changes to the group's structure?

Gaming The System

To return to our climate change formulation of the prisoner's dilemma, the best overall outcome might indeed be for each individual to sacrifice some immediate comfort to curb total carbon emissions. But if you're convinced that everyone else will sacrifice, it doesn't make sense for you to do so as well – after all, your actions have such a small global impact, and you don't want to give up immediate comfort. If everyone else *won't* sacrifice, then you definitely don't want to give up immediate comfort, since you cannot singlehandedly stop climate change.

This logic drives the natural human tendency to "free ride" on the contributions of others. Not as much as theory would predict, to be sure – we have a great capacity to be altruistic, hard-working, and cooperative. But we are also prone to follow the path of least resistance. The solution to a real-world prisoner's dilemma, then, is not to moralize and hope, but to change the very payoff structure to induce cooperation.

In public policy, this might mean taking a page from the VCM research we discussed and compelling people to contribute their fair share. If you're a mob boss, this might mean offering your crew a large bonus if they refuse to rat on their fellow criminals (or, more realistically, punishing those who do rat). In any case, understanding the prisoner's dilemma means understanding how to change the game to *avoid* playing the prisoner's dilemma.

In the next chapter, we'll continue our discussion of the link between game theory and the real world by analyzing how more lessons from the field can be put into practice.

Key Takeaways

- The prisoner's dilemma is an important game in game theory that describes a situation in which acting selfishly is the dominant strategy for an individual, even though acting cooperatively would lead to a social optimum. Elements of the prisoner's dilemma can usually be found when selfishness and cooperation are in tension.
- The prisoner's dilemma was formalized and refined at the RAND corporation, a US government-sponsored think tank, in the aftermath of World War II. The ethical issues resulting from the dilemma, though, have been studied throughout history, including in ancient Greek philosophy, Christianity, and social contract theory.
- A dominant strategy is a strategy that is best for you to play regardless of what your opponent does. A Nash equilibrium describes a state in which neither player has an incentive to change strategy, given that their opponent's strategy stays

fixed. A Nash equilibrium can differ from a social optimum, which is a state where total social welfare is maximized.
- While game theory predicts a cooperation rate in the prisoner's dilemma of zero, experimental cooperation rates can be greater than 40%. Additionally, significant cooperation can be a high-performing strategy in repeated versions of prisoner's dilemma, including in the simple but effective 'tit for tat' approach.

3

THE HIDDEN CHESSBOARD - HOW GAME THEORY SHAPES OUR WORLD

Viewed in retrospect, the founding of Apple seems like destiny. As a visionary entrepreneur, Steve Jobs saw a world no one else could, a world where high-powered computers could sit in your pocket and connect you to every person on earth. This story is commonly understood to be one of an unrelenting genius setting his sights on greatness, making the resulting achievements seem almost inevitable.

Although this narrative has inspired not one, but *two* Hollywood biopics, the real story is not quite so simple. While Jobs was certainly a person of unusual drive and tenacity, starting Apple with his friend Steve Wozniak was a gamble based on prevailing trends, not a decision informed by a clear vision of the future. In fact, Jobs' original passion was Zen Buddhism. Before starting Apple, Jobs spent seven months in India seeking enlightenment, followed by a brief stint working at a farming commune in Oregon.[1]

Nassim Taleb, a former trader and author of a series of books on uncertainty and probability, dubs the manipulation of ambiguous events into a clear-cut story the 'narrative fallacy.'[2] We live our lives forward, but can only understand them backward, leading to our penchant for compressing contingent events into a series of logical conclusions.

To better understand the story of Steve Jobs, and avoid the narrative fallacy, we can use the tools of game theory to explore the young entrepreneur's decision-making. In fact, game theory can be used to analyze many aspects of competitive decision-making in the real world. In this chapter, we'll see how we can apply the field to business, technology, and politics in order to better understand the world around us.

Apple Seeds

Suppose you are a young Steve Jobs, trying to figure out your path in life among different competing options. While you have a strong passion for meditation and Zen Buddhism, you also see opportunities in the burgeoning field of personal computing. This option could be risky, though – if personal computing turns out to be a fad because of consumer indifference or a lack of business investment, you might spend many years working in the field only to wind up broke.

On the other hand, with your charismatic personality and deep personal experience, you believe you could be successful in continuing down the path toward enlightenment and eventually becoming an influential teacher. The payoff here, though, would arguably be smaller than starting a successful computer company.

If other people continue to invest time and resources into the personal computing space, then you should pursue opportunities there. If they don't, however, your best bet is to follow your original passion in Buddhism and meditation. In other words, you have to make a choice whose results depend on the actions of other people.

This dynamic, through which entrepreneurs and businesses decide which sectors to enter based on what they expect other people to do, results in the formation of many different types of market structures. Each structure has its own idiosyncrasies, but all can be understood through the lens of strategic decision-making and game theory.

The classic perspective on decision-making in business is that a firm should seek to maximize its individual profit assuming that the market is stable. In the real world, though, the market is certainly not stable. The game theory perspective expands the classic view to understand how a firm's decisions will change the market as a whole. In game theory, unlike in traditional economic analysis, the decisions of and impacts on competitors must be considered. Since these decisions and impacts will come back to affect the profitability of the original firm, the game theory perspective can lead to better choices by business management.

Perfect Competition

We can begin our discussion of game theory with respect to market structures by analyzing the classic structure studied by basic economics courses: perfect competition. Perfectly competitive markets are, as it turns out, the least interesting markets to study with a game-theoretic lens, but the reasons why will motivate our study of more exciting market structures.

Perfectly competitive markets are characterized by the presence of sufficient firms selling identical products that none of them have any pricing power. In addition, there are enough consumers in the market purchasing the products that no consumer has any pricing power. Therefore, the market price is taken as given, set by the supply and demand of the market as a whole, not by any individual buyer or seller. The market for blueberries, and other identical agricultural products, best resembles perfect competition in the real world.

From the perspective of firms, there are essentially no interesting strategic choices to make in a competitive market. The analytical result of a competitive market is that any firm selling their products for a price lower than the current market price would quickly go out of business, as they would be selling for a loss. Any firm selling for a price higher than the market price, of course, would not sell anything, since consumers would buy the product from another firm. Firms merely try to maximize their accounting profit, which is just enough to make business worthwhile.

Monopolistic Competition

Markets exist on a spectrum between the perfectly competitive and the pure monopoly. A monopolistic market shares many traits with a perfectly competitive market but is slightly farther along the spectrum toward monopoly.

In a monopolistic market, there are many firms competing, but each firm sells a differentiated version of the product. A classic example is the restaurant business. Restaurants compete in the market for prepared food, but a meal at a French restaurant is not a perfect substitute for fast food (for that matter, two French restaurants might not be perfect substitutes for each other, depending on how picky you are about your *steak frites*).

Strategic decision-making in monopolistic markets mainly concerns marketing decisions, and, by extension, market segmentation decisions. For instance, if you are selling beer, you might focus on cultivating a brand that connects with working-class people, especially if you believe your competitors will target wealthier demographics. While you might be selling for a lower price point in this market segment, your domination of the segment could lead to a higher profit overall.

Similarly, your decision about the amount of money to spend on marketing overall will depend on your competitor's choices. If the beer industry currently has low product differentiation, but also low advertising expenses, the industry could be in a stable equilibrium. If one firm begins to differentiate through advertising, though, it will likely force other firms to spend more on marketing to keep pace. This strategic interaction, where any one firm must consider how its choices will impact competitors, is a hallmark of game theory analysis.

Oligopolies

As we move further away from competitive markets and towards a monopoly market, we approach the market structure known as oligopoly. In an oligopoly, there are so few firms in the market that each one has significant pricing power. In extreme cases, there can be just two firms in an oligopoly, a situation known as 'duopoly.' Duopolies tend to appear in markets with high barriers to entry, such as the airplane manufacturing market, dominated by Boeing and Airbus.

Oligopolies are arguably the most common type of market in the modern world, with many major industries being dominated by just a few firms. Oligopolies are particularly interesting from a game theory perspective, because firms have significant pricing power, and therefore have to make strategic decisions about both price and non-price competition.

With respect to non-price competition, firms operating in an oligopoly have to make similar strategic decisions to firms operating in a monopolistic market. Advertising budgets, product differentiation, brand identity, ease of use, and market segmentation decisions all influence an oligopolistic firm's profits and pricing power.

It is in price competition, though, that firms have to incorporate game theory the most, since their pricing decisions can heavily influence competitors. For instance, if both Boeing and Airbus keep their prices high, the airplane manufacturing market as a whole could likely achieve the highest *total* profit. But if one firm cuts its price, it might capture more of the market, and therefore earn a higher *individual* profit. This would incentivize the competing firm to cut prices as well, until the industry stabilizes in a low-profit equilibrium.

This structure creates some obvious incentives for the two firms, of course. If Boeing and Airbus could find some way to cooperate, ensuring both firms keep their prices high, they could earn a higher profit than if they succumb to price competition. This type of collusion is illegal in most developed countries since it harms consumers by artificially raising prices. In the 1990s, five companies were prosecuted by the United States Department of Justice for conspiring to raise the price of lysine, an animal feed additive.[3]

Although explicit collusion may be illegal, oligopolistic firms can implicitly collude under the guise of "price leadership." Under price leadership, a dominant firm will set a price, with the rest of the industry clustering around that price as a focal point. While there may not be explicit communication among the firms to fix prices, the result is essentially the same.

In fact, two German economists studying the Italian gasoline market used statistical analysis to determine that the effects of tacit collusion under a price leadership dynamic could be as powerful as explicit collusion. In their research paper, the economists analyzed how a new pricing strategy from market leader ENI resulted in other firms adjusting their pricing strategy as well. The end result was higher prices for Italian consumers, both compared to the pre-adjustment period and when compared to Europe as a whole, despite no evidence of explicit collusion.[4]

Monopolies

Monopolies are somewhat rare in the real world, because they are typically easy to identify, and competition law is primarily written with the aim of avoiding market domination by a single firm. Regulated monopolies, like public utilities, do exist, but they have very limited strategic decision-making due to the nature of public oversight. Still, game theory can teach us some unintuitive lessons about monopolies, with extensions beyond market structure analysis.

The commonly understood justification for why monopolies are bad for society is that monopoly firms artificially restrict supply in order to keep prices high. By doing so, they maximize their profit in a way that would not be achievable given adequate competition.

What this analysis misses, though, is that it is not enough merely for a monopoly to *begin* at a high price; they must *commit* to keeping prices high in order to reap their rewards. Consider that a monopoly always has an incentive to sell additional goods so long as the price exceeds the cost of producing those goods. For instance, suppose a monopoly sets the price of some product at $20. To maximize profits, once the monopoly sells the product to all the customers willing to pay $20, it can just drop the price by a dollar to sell to more customers who would pay $19. This process will continue until the monopoly ends up selling at a price equal to its cost of production.

Of course, given this logic, all consumers would want to be part of the last cohort to buy goods, since this group will get the best price. Therefore, a monopoly firm would not sell *any* goods at the original, artificially high price (except to the most impatient of consumers).

One way to break this logical chain is for the monopoly to commit to not lowering the price, meaning consumers would have no incentive to wait to buy from the firm. This only works, of course, so long as consumers believe the firm will keep prices high, which might be achieved by the firm restricting (or even destroying) its own manufacturing capacity.

In business, as in life, a commitment is only as powerful as it is credible.

Game Theory In Technology

While game theory is important in the strategic decision-making of technology firms, the field has also found surprising applications in the development of technology itself. Specifically, computer scientists have been able to leverage game theory to research challenging optimization problems.

Braess' paradox, for instance, is a result in the field of traffic design that represents a multiplayer prisoner's dilemma. The paradox is that adding an additional shortcut to a road network, which speeds up the route for an individual driver, can actually slow down the total travel time for drivers taken as a whole. In other words, the individually optimal Nash equilibrium is not necessarily the socially optimal outcome.

The paradox is not necessarily a curious idiosyncrasy, either. In 1983, two researchers from Columbia University and the University of Chicago published a theoretical paper detailing how, under fairly general assumptions, the paradox is about equally likely to occur or not occur.[5] The result has important implications not just for traffic design, but for any congested network, including the computer networks the internet relies on.

Game theory has also played a role in recent developments in artificial intelligence (AI), particularly in reinforcement learning. When AI models are being trained, they are typically attempting to optimize some game-theoretic structure, in which an initial input is used to make a decision that leads to a reward. Through a trial-and-error process, these models learn how to optimize their strategy for a given game. This link between AI and game theory also provides context for why AI models struggle when attempting tasks that cannot be simply modeled as a game, such as driving.[6]

Game Theory In Politics

A survey of game theory's applications to the real world, of course, would not be complete without a discussion of game theory and politics. As a situation with competing interests and different potential strategies, politics can be understood as a game, though its potential impacts make it one of the most high-stakes games discussed so far.

Strategic voting is one way game theory manifests itself in politics. Since elections are zero-sum games, with just one winner and one loser, voters will compromise their true preferences based on what they expect other voters to do. If a voter has a strong preference for a candidate unlikely to win an election, they have an incentive to vote instead for the candidate with a potential chance of winning that they can most tolerate. This problem plagues third-party representation in American politics, where voters are seen as 'throwing away' their vote if they do not vote for either of the two major parties. Alternative voting systems, such as ranked-choice voting, have been proposed to minimize the use of strategy in voting and allow people to express their true preferences.

Once politicians are elected, of course, the use of game theory does not end. Political negotiations are strategic affairs, with each side trying to gain the upper hand. As this book is being written, debt ceiling negotiations are underway in America, in a situation that closely resembles the chicken game described in chapter one. Neither party wants the potentially disastrous consequences of a default to come to fruition, but each side wants to credibly threaten such an event if they don't get their way. Proposals for creative legal solutions to bypass the negotiations are an admission that the best way to win some games is to avoid playing them altogether.

Action Steps

Clearly, game theory is not just a subject of academic interest, but a field that can have real-world consequences. Answer the following questions to deepen your understanding of the practical impacts of game theory.

1. While monopolistic markets and perfectly competitive markets are similar, they differ in the fact that firms in monopolistic markets produce differentiated, not identical, products. How does this small change impact the extent to which firms in either market can use game theory?

2. The following payoff structure represents potential pricing strategies by Uber and Lyft, who are competitors in the oligopolistic ride-hailing market. What game, described in a previous chapter, does this structure most closely resemble? Do the firms have a dominant strategy?

<u>Ride-hailing Market</u>

(Uber, Lyft)	High	Low
High	(100, 100)	(50, 120)
Low	(120, 50)	(80, 80)

3. Remain with the ride-hailing market payoff structure. Suppose you are an industry consultant advising both firms in a country without competitive regulation. Describe a proposal you might make to the firms so that they can maximize their total profit, including enforcement mechanisms. If the firms reject your proposal, what other possible market dynamics might result?

4. What are the positives and negatives of politicians becoming more well-versed in game theory? Do you think politicians in your country should understand game theory better, in order to lead to more efficient problem-solving, or would it tend to make them more ruthless in pursuing their objectives?

Steve's Stag Hunt

At the beginning of the chapter, we told the story of Steve Jobs and the difficult decision he had when weighing his future career paths. Now that we have experience applying game theory to the real world, let's take this dilemma and find a game that best matches the situation.

As we discussed, choosing to start Apple and work in personal computing will only be successful if other people sustain the growth of the industry. On the other hand, if Steve follows his passion for Buddhism and meditative teaching, he is less reliant on the choices of other people, but his potential success is much more limited.

This decision closely mirrors a game known as the stag hunt (also known by less colorful names such as the trust dilemma or the common interest game). The stag hunt was originally formulated by Swiss philosopher Jean-Jacques Rousseau, who described a situation in which two hunters could either work together to catch a large stag, or hunt individually to catch small hares.[7] While the stag has a larger payoff, the hunters will only be successful if they *both* choose to hunt the stag.

The payoff structure for hunters A and B, looks like this.

Stag hunt

(A, B)	Stag	Hare
Stag	(10, 10)	(0, 5)
Hare	(5, 0)	(5, 5)

If hunter A chooses to hunt the stag, then hunter B should also choose the stag. But if A chooses hare, then B should also choose hare – as B has a payoff of zero if they waste their time hunting the stag without the other hunter. Similarly, starting Apple was only the optimal choice for Steve Jobs if he believed personal computing would continue to grow. If consumers and investors quickly moved on from personal computing, he would have been wasting his time when he could have pursued other interests. But if he pursued other interests, he would have been wasting his time if personal computing took off.

The stag hunt emphasizes a fundamental truth of strategic decision-making, and one that game theory relies on: no decision is made in a vacuum. The results of our choices depend on the choices made by other people. In the next chapter, we'll use this idea and the other lessons of game theory to evaluate our own, personal decision-making process.

Key Takeaways

- In a perfectly competitive market, firms have little use for game theory because there is essentially no strategic decision-making to be done. In the real world, though, perfectly competitive markets are the exception, not the rule.

- In a monopolistic market, most strategic decision-making revolves around marketing. Products in a monopolistic market are primarily differentiated through branding, although decisions about product quality and service location also need to be made.
- In an oligopoly, firms have significant pricing power as a result of being one of the few market participants. This gives them ample room to use game theory to inform pricing decisions. Collusion can often result in higher industry profits overall, but there is usually an incentive to defect from collusion in search of higher individual profits. Tacit collusion, while generally not illegal, can have similar impacts as explicit collusion.
- Monopolies generally do not have strategic decisions to make, as they simply set a price to maximize profit, but they may have to commit to keeping prices high in order to ensure sales. If customers thought they would eventually lower profits after exhausting all sales at the original price, they might simply wait to be part of the last group to purchase goods.
- Game theory has found significant applications in technology and computer science research, particularly in the study of optimization and equilibrium problems, which can often be modeled as a multiplayer game. Game theory is also used in artificial intelligence, where models

are often trained under a game-theoretic structure.
- Politics, which features competing interests using strategy to pursue specific outcomes, is a field where game theory can frequently be applied. Citizens often use strategic voting to make their votes count the most, while politicians use binding threats and intentionally vague messaging to achieve their goals in negotiations.

4
MAKING GOOD CHOICES - WHAT YOU CAN LEARN FROM THE KINDERGARTEN PLAYGROUND

Do toddlers strategize, or is long-term planning only for adults?

Consider the case of Kate and Emma, two young children playing on their kindergarten playground. Kate is happily using a set of plastic tools to mimic building and destroying different structures. Emma wanders over and grabs one of the tools, looking to join in on the game.

There is a brief moment of tension as Kate sees someone else using her toys. Eventually, though, Kate decides to let Emma share the tools, and the two go on playing together.

From an outsider's perspective, it might seem that some kindergarten teacher's admonishments to *share* have finally had an effect. From this point of view, Kate is learning to be nice, shedding the selfishness that nearly all young children find natural.

As game theorists, though, we might see an element of strategy in Kate's choice to share. Perhaps Kate has finally developed enough maturity to consider the impact her decisions *today* will have on her life *tomorrow*. That is, if Kate shares her toys with Emma now, maybe Emma will be more willing to share with Kate in the future.

This story of Kate's decision is a bit more pessimistic, since it portrays cooperation as motivated by selfishness. Ultimately, which story you think is more likely depends on your views of early childhood development. Still, the situation teaches us an important lesson as we continue to explore the field of game theory: even the smallest, most intuitive decisions contain elements of strategy. Simply put, everyone is a strategist.

In this chapter, we'll synthesize the knowledge we've gained in previous chapters to reflect on our own decision-making process. We'll explore how the tools of game theory, including thinking strategically and considering other people's choices, can improve the way we make decisions. Although we are likely to have many more high-stake choices than whether or not to share toys on the playground, the example shows how strategic thinking can lead to better individual and social outcomes.

Before diving into the lessons learned from the games we've discussed so far, it's important to remember the structural differences between clearly defined games and real-world situations. As it turns out, though, these differences can teach us something important about avoiding indecisiveness and uncertainty.

The Trouble With Boundless Freedom

One of game theory's main limitations when applied to the real world is the lack of clearly defined strategies in life. In game theory, competitors typically choose between one of two paths, which are executed exactly the way the game defines. In real life, the possible decisions that could be made are endless, and each general route offers an infinite number of potential diverging branches forward.

For instance, suppose you and your spouse are trying to decide between going on a hike or going to a baseball game. In theory, this sounds like a simple choice based on the respective enjoyment you and your spouse would experience by attending each event together and the enjoyment of attending each event separately. In reality, of course, this decision will depend on an endless number of variables. What hike will you choose to do? Will you drive to the game or take public transit? What snacks will you bring on the hike? Do you plan on inviting friends to the game or not?

In the real world, there is simply no way to list every strategic option, since each option has endless variation. Similarly, the results of the option you choose will depend on factors outside the scope of your strategic thinking. In the American sitcom *The Office*, branch manager Michael Scott enlists a number of tactics to gain the upper hand during salary negotiations with an employee, but embarrasses himself by accidentally wearing a woman's suit to the meeting. This is hardly a factor Michael would have included in a strategy analysis.

Just like a game theory setup would be untenable with millions of strategic options, making choices in life would be impossible if we did not take steps to reduce the complexity in our decision-making process. This idea is closely related to the paradox of choice, a key concept in decision-making research.

The paradox of choice describes how giving people more choices can, counterintuitively, make them less likely to make a decision. In one famous experiment, researchers reduced the number of jars of jam on display in a supermarket from 24 to 6, which increased the number of jars purchased by more than seven times.[1]

Game theory can teach us an important lesson to reduce the indecisiveness inherent in being presented with endless optionality: When presented with many potential paths, decide on several well-defined strategies to actually choose between, while ignoring the decisions that won't have a great impact. As the saying goes, any decision is better than no decision at all. In real life, avoiding indecisiveness can be just as important as avoiding the wrong choice.

Of course, once we've clearly defined our choices, we still have to play the game. Thankfully, the lessons of game theory don't stop at merely framing our decisions. As it turns out, one of the most useful lessons from game theory, especially when it comes to navigating real decision-making, is also one of the field's most unexpected lessons.

Walk A Mile In My Shoes...

Game theory is typically framed in a very competitive way. This is perhaps unavoidable in a field explicitly designed to study optimal decision-making among participants with diverging interests. But despite this focus on individual competitiveness, one of the overarching lessons of game theory is centered on a concept that stands in contrast to competition: empathy.

The importance of empathy, or the art of putting yourself in another person's shoes, might seem a strange lesson to draw from game theory. In truth, though, the only way to strategize is to consider other people's perspectives. In the prisoner's dilemma, as introduced in chapter two, one prisoner can understand the other's likely strategy by viewing the situation as their accomplice would. It is not by luck or premonition that each captor understands what the other is likely to do, but by empathy.

In its simplest form, using empathy strategically can mean understanding that considering yourself to be the sole decision-maker in a situation is unlikely to yield the best results. For instance, if you are driving home from a sporting event, you might be tempted to rush out the door at the end of the game in order to beat the traffic home. But if you consider that many other fans are likely making the same decision, you might choose to hang around for a while after the game ends, saving yourself from sitting in the initial rush of traffic away from the stadium.

In tune with the common understanding of empathy, of course, being empathetic can also lead to great cooperative benefits. While empathy can improve individual outcomes, it can also lead to better social outcomes by forcing decision-makers to consider what their opponents want. In doing so, decision-makers might discover unknown areas of potential mutual benefit.

For instance, suppose you are frustrated with a micromanaging boss, who simply cannot leave you alone to get your work done. While your initial reaction might be to scheme about the best ways to structure your schedule to avoid being in the office at the same time as your boss, you might also benefit by considering the situation from their perspective. Perhaps they are simply anxious about their lack of control over certain projects. If this is the case, a simple status update email at the end of each day might go a long way to help reduce stress for both parties.

Empathy can also keep us from the worst implications of game theory's rationality. Many of the initial founders of the field, including the eminent John von Neumann, were advocates for America conducting a preemptive nuclear strike on the Soviet Union during the Cold War.[2] In their view, the clear logic of game theory meant that nuclear war was a mathematical certainty, so the only way to ensure victory was by being the first to attack.

Thankfully, such an outcome did not occur, precisely because real games are played by human beings. As humans, we can empathize with our opponent's hopes and fears – in other words, the irrationality that makes us human. Whether it was the dread involved in being the one to press the red button, or the leap of faith involved in trusting your enemy to act as you do, empathy trumped the advice of game theorists during those fateful decades.

When making decisions, we can use empathy not only to consider other stakeholders as strategic decision-makers, but also as human beings, just like us. Empathy is certainly important for the way it can grant an individual a competitive advantage, but also for the way it opens up opportunities for cooperation and collaboration.

While empathy is tremendously valuable, soft skills are not all that's needed for optimal decision-making. Sometimes, making good choices requires a little fact-finding.

Just The Facts, Ma'am

In the 1987 film *Wall Street*, corporate raider Gordon Gekko (played by Michael Douglas) declares that information is the most valuable commodity he knows.

As someone well-versed in the art of strategy, Gekko is intimately familiar with one of the most important lessons from game theory: the power of information. While framing choices properly and considering your opponent's perspective will certainly help you make better choices, they cannot overcome limitations in your own knowledge.

Often, the importance of information in game theory is obscured by the idealized set-ups, where all information is provided. As game theory research needs to be focused on tractable problems in order to make progress, academic research tends to simply assume that payoffs, participants, strategies, and situations are all known with exact certainty. In real life, of course, this could not be farther from the truth.

Suppose you are a manager with many direct reports, and are interested in structuring your team's environment to ensure employees are happy, healthy, and motivated. Before investing in catered lunches or renovating the office, you might find tremendous benefit in slowing down and investigating what truly motivates each of your individual employees.

For some, it might be a more flexible working arrangement to allow them to spend more time with family. For others, knowing that a raise sits on the other side of a completed project might be all the motivation they need. Certain people can only find true motivation in work they are intrinsically interested in, so the best results might occur from letting them pick their own projects.

Whatever the case, this type of critical information cannot simply be intuited or assumed. Whether in career choices, business negotiations, or even family decisions, knowledge is power. Making a decision without ascertaining all the relevant facts is a surefire way to regret the strategic choice you made, resulting in a waste of time, money, or both.

Moreover, gathering more information can actually make it more likely that you are willing to make a decision, similar to the way that clearly defining choices can reduce indecisiveness. Ample experimental research indicates that people are more willing to take risks when they gather more information, even if their beliefs about the risks haven't changed at all.[3]

For instance, suppose you have an uninformed belief that a new business you want to start has a 60% chance of success. After conducting weeks of market research, interviewing potential customers, and talking to investors, your belief remains the same – a 60% chance of success. The power of information is such that even though your perception of the risk has not changed at all, research indicates that you will be more likely to start the business than you were previously.

Sometimes, in searching for the important facts, you'll stumble on a piece of information that narrows down your strategic options to one clear choice. Such situations are rare, but when they appear, game theory can prepare you to take full advantage.

Dominant Strategies

After empathizing, fact-finding, and clearly defining your choices, you may realize that one strategic option stands out above the rest. Such a strategy is called a "dominant strategy" for the way it dominates all other choices.

We spoke briefly about dominant strategies in our discussion of the prisoner's dilemma in chapter two. Recall that in the prisoner's dilemma, each prisoner has a dominant strategy to testify, regardless of what their accomplice does. Whether their accomplice stays silent or not, each prisoner can only improve their situation by testifying, so testifying is a dominant strategy.

Dominant strategies often result in business when one party in a negotiation has little to lose by walking away. For instance, suppose you are discussing salary for a new job opportunity. If you have other opportunities lined up, pushing for a higher salary is likely to be a dominant strategy for you. If the company declines, you can walk away and try your luck at the next company. If they accept, all the better. It is only when you have few remaining opportunities that holding out for more money no longer dominates accepting what you've been offered.

Dominant strategies can also result from the existence of strong personal preferences. Imagine you and a friend were deciding whether to eat dinner at an Italian restaurant or a Spanish restaurant, but the two of you never did make a decision. Now, dinner time is approaching, but your phone is dead and you have no way to contact your friend.

If you really enjoy *tapas*, heading to the Spanish restaurant is likely to be your dominant strategy. Either your friend also goes there, and you enjoy a meal together, or you eat delicious food by yourself. Ending up at the Italian restaurant alone, however, would be the worst scenario. Note that if your friend also knows that you love Spanish food, the strategy is even better, since they will assign a very high probability to your going to that restaurant, and are likely to head there as well.

Taking a field as academic as game theory and applying it to your daily life is not without challenge, but discovering dominant strategies, which could lead to major benefits in our personal and professional lives, is likely to make the work worth it.

Action Steps

Now that we've explored the key practical lessons from game theory, test your understanding by answering the following questions.

1. Suppose you are thinking about career options after completing your college degree. Knowing that the overwhelming number of options before can be debilitating to decision-making, what steps might you take to simplify your strategic choices?

2. Suppose a national leader is summoning her closest advisers during a time of war. Is it possible that her military adviser and humanitarian adviser will both tell her about the importance of empathy? If so, in what ways might their advice be similar or different?

3. While the value of information is tremendous, some would argue that we live in a world with far too much access to information, especially via social media and the internet. As game theorists, how can we determine what information we need to pay attention to? Is it possible to research or explore too much information when it comes to decision-making?

4. Consider the story about you going to dinner with a friend in the section about dominant strategies. Could your dominant strategy change if your friend loves Italian food? What about if you are indifferent between eating alone or with your friend? Create a payoff matrix that results in a dominant strategy of going to the Spanish restaurant, and one that does not.

It's A Sign

As we've seen in the last few chapters, game theory can teach us lessons not only for the classroom, but also for the real world. Since our days are filled with strategic choices that influence our lives, it's imperative that we clearly define our choices, remember the way our actions impact others, and, perhaps most of all, gather information to make intelligent choices.

Sometimes, though, information is just too costly to obtain. Since our days are filled with choices, we cannot spend all of our team researching every single aspect of our decisions. To that end, we often use shortcuts to skip the expense involved in gathering true information.

Let's return to Kate and Emma, the two kindergartners playing on the playground. Suppose a father who was interested in putting their child in that kindergarten was touring the school that day and saw the playground episode. Seeing the two kids sharing toys is likely to be a positive sign for the father, as he sees that the kindergarten teachers are capable of inculcating children with a sense of kindness and responsibility.

In other words, the father uses the children's actions to judge the teachers' competency, information that is very difficult to obtain without observing lessons for days on end. Although the signal may turn out to be wrong, relying on a signal for certain information is often vastly more efficient than finding the information itself. This mental shortcut helps us navigate the complexity of the world around us.

In the next chapter, we'll discuss the informative content of signals, and what role signaling plays in our decision-making process. We'll also explore situations in which this mental hack goes awry, and why we should always be careful of relying too much on what we think we know.

Key Takeaways

- Game theory teaches us to clearly define several competing options to make decisions more effectively and avoid the paradox of choice.
- By using empathy and viewing situations from our opponent's perspective, we can identify their next likely move and adjust our choices accordingly.
- Without knowing the facts, we risk making the wrong decision. Gathering information is vital to being a good strategist, a fact often underappreciated in game theory, since academic research often assumes all information is known so that calculations are more tractable.
- Sometimes, strategic thinking can identify a single path forward that is better than all other options. Such a path is known as a dominant strategy, and it is vital that we take advantage of it if we identify one. Dominant strategies can be rare, however, so it is equally important you check that you are fully informed about the situation.

5

SIGN LANGUAGE - WHEN TO FIT IN AND WHEN TO STAND OUT

In the Hebrew Bible, a famous story is told about the aftermath of a battle between two tribes, the Gileadites and the Ephraimites. The Gileadites, having defeated the Ephraimites, wanted to stop any enemy fighters from crossing the Jordan River back into Ephraim. The two tribes had no clear distinct physical characteristics, however, so the Gileadites could not easily distinguish friend from foe.

The Gileadites found a creative solution that depended on the way each tribe spoke. They asked anyone trying to cross the river to say the word 'Shibboleth,' knowing that the Ephraimites could not pronounce it the same way the Gileadites could. Anyone who failed the verbal challenge was killed on the spot.[1]

This practice, of outing the enemy through linguistic quirks, continues in the modern era. In World War II, Allied forces on D-Day would challenge unknown combatants with the phrase sequence 'Flash,' 'Thunder,' and 'Welcome,' due to the difficulty German speakers had pronouncing the words.[2] For a less martial example, consider how New Yorkers can easily identify out-of-towners by asking them to pronounce 'Houston Street.'

The term 'shibboleth' has now come to mean any custom that can be used to differentiate between group members and outsiders. Shibboleths are a fascinating real-life manifestation of the subject of this chapter: signaling and screening. Signals are present in nearly every aspect of life, from our social interactions to our business affairs, and help to simplify the information-gathering process. They serve as a form of communication meant to convey information about some underlying fact. These signals are screened by receivers to judge their true information content.

To explore this concept in depth, we'll begin with a description of the economic theory that underlies signaling and screening in strategic behavior, followed by an exploration of how signals appear in various aspects of life.

Signaling And Screening

In chapter four, we described the importance of gathering information when it comes to making decisions. As we alluded to in the conclusion of the chapter, gathering accurate information can be an enormously difficult and expensive undertaking. This is especially true when trying to establish things that are genuinely unobservable, like a person's moral character or social standing.

Instead of being investigated at the source, the majority of our information is actually gathered through the use of signals and screening. By interpreting some factor that is cheap to observe, like the way someone pronounces a word, we attempt to deduce information that may otherwise be costly to obtain, like a person's true place of origin. The initial signal is subject to the secondary act of screening, from which informative content is drawn.

In game theory, and strategic behavior in general, signals are important for the way they convey information about a person's likely future behavior. In chapter one, we described how a participant in a bar dispute could communicate a willingness to fight by wearing a t-shirt from a local boxing gym. Using our new terminology, this t-shirt would act as a signal that an opponent would have to screen to ascertain information; in this case, the likelihood of a dispute escalating to violence.

Signaling exists whenever some behavior or choice is meant not only to achieve an end in itself, but also to serve as an element of communication. Of course, this aspect of signaling means that clever strategists can leverage certain signals to communicate things that are not entirely true. For instance, consider a man who, on a first date with a woman, brings her flowers, opens her car door, and pays for dinner. While such behavior may mean the man is truly a gentleman, he may also simply want to convey the appearance of being a gentleman for his own ends.

This inherent suspicion, that people are faking signals in order to get us to believe certain things, is why we pay far more attention to signals that incur significant investment. As the saying goes, talk is cheap – speech is a cheap signal compared with genuine action. Nowhere is this more evident than in the importance of college degrees in signaling competency for employment.

Earning a degree is an arduous process, involving many hours of study over the course of several years. For this reason, earning a degree is a stronger signal for someone's competency than someone merely claiming to be competent. As it turns out, the signaling power of a degree is so powerful that, for job prospects, it is far more important than one's actual education, even though the degree is ostensibly meant to represent such education.

Can Colleges Be Disrupted?

Over the past several decades, as the rise of the internet has made it easier and easier to cheaply access the collective knowledge of humanity, prognostications about the eventual collapse of the university system have been made. As the prediction goes, in a world where people can learn almost anything for free online, there would be no need to pay exorbitant tuition fees for a traditional college education. Yet, in 2023, the university system seems as strong as ever. Elite institutions continue to receive a huge volume of applications from eager high school students, despite the knowledge available cheaply online.

Such mispredictions likely have to do with a misunderstanding of the purpose of a college education. In truth, what is important about college is not the education that it entails, but the degree that it grants successful students. While earning an education via self-directed research or open online courses is possible, there is no degree at the end of such a process. It is the college degree, which signals intelligence, diligence, and reliability to prospective employers, that justifies college tuition.

A famous research study, published by two University of Michigan economists in 1978, found compelling evidence for what is known as the "sheepskin effect," or the tendency for an earnings premium to be associated with a degree rather than the underlying education it is meant to represent. [3] The economists analyzed population data and found that a significant jump in expected wages was observed after people completed 16 years of education (the general amount it takes to earn a college degree). If the value of college was in the education, and not the degree, then one should expect a smooth increase in earnings with each additional year of education instead.

Recently, economist Bryan Caplan provided supporting evidence for this approach in his 2018 book *The Case Against Education*.[4] While this argument is not particularly popular with educators, who believe in the intrinsic value provided by a college education, it does provide a convincing explanation for why the university system is here to stay. The internet is a great resource for education, but education is not why college is valuable.

A college degree, by aligning with the traditional pipeline for professional success, also tends to signal a certain conformity to social standards. In a similar way, certain performative behaviors by career professionals, which at first glance may seem inefficient or silly, make more sense when viewed from a signaling perspective.

Suits And Sparrows

If you've worked in an office before, you may have encountered the situation of needing to dress a bit more sharply due to clients coming in for a meeting. If your standard office wear is on the casual side, it might be seen as unprofessional to wear such clothes to a client meeting. For men, a suit might be required, although the need for a tie seems to have disappeared over the past decade.

From one point of view, this is patently absurd. Aside from the slight shift in mindset that might be associated with donning a suit, wearing nicer clothes cannot possibly have any influence on the quality of your work, which is what clients should really care about. Why, then, does this tradition seem to be so common?

With our new signaling framework, we can understand the suit as a signal of professionalism and care. In certain fields, the quality of work can be enormously difficult to determine in real time, particularly in consulting and finance where professionals are paid for advice. Professional clothing and demeanor can communicate an air of responsibility and conscientiousness to a client, which is a helpful signal in the absence of clear information about work quality.

Interestingly, similar forms of signaling appear to take place in the natural world. Certain sparrows, depending on their position in the dominance hierarchy, are known to develop darker plumage on their bodies. This darker plumage communicates to competitors that the dominant sparrow is not to be trifled with. Apparently, this signal is so strong that sparrows with naturally darker plumage are rarely ever challenged to prove their dominance.[5]

So, when considering your demeanor or presentation at work, remember the sparrow. Sometimes, to convince others of your skill, signaling that you have the skill is more important than actually having the skill.

In certain situations, though, this logic is turned on its head. Sometimes, refusing to signal certain traits can actually provide the most information of all.

Countersignaling

In his 2018 book *Skin in the Game*, author Nassim Taleb discusses at length an extension of traditional signaling theory known as "countersignaling."[6] Countersignaling is the act of inverting traditional signals in order to, paradoxically, send a stronger signal about one's characteristics than would otherwise be possible.

For instance, imagine that you suffer from a serious medical condition, and have to select a doctor to perform a risky surgery. The first doctor is the image of professional success in the medical field; white coat, smiling face, and multiple Ivy League degrees. The second doctor, while equally as successful as the first doctor, speaks with a Brooklyn accent, has a degree from a less-respected institution, and seems more interested in interrogating you about your diet than having good bedside manners. Which doctor should you choose?

Taleb argues that, contrary to our intuitions, we should jump at the chance to work with the second doctor. His reasoning is that a doctor that does not fit the traditional mold of a doctor must be far more skilled to achieve such a level of success. Doctors who look like doctors, on the other hand, might be able to skate by on the power of their signals rather than their skills.

Similarly, some evidence from the financial services industry indicates that investment firms run by women tend to outperform those that are run by men. This makes sense if we consider that women do not fit the traditional mold of successful investment manager, since men dominate the industry. Therefore, to convince investors to allocate capital to their firm, women will have to be more skilled than the industry average.

Countersignaling also manifests in politics, where clever politicians utilize the concept to convey the appearance of not being a typical politician. During the 2020 American presidential election, the Democratic primary debates were chaotic affairs, with ten candidates on stage all vying to stand out. Andrew Yang, likely in an attempt to showcase his tech-oriented mindset and willingness to be different than traditional politicians, went tieless on stage, the only man to do so.[7] While Yang's campaign was ultimately unsuccessful, the countersignal certainly got people talking, and helped him stand out in a crowded field.

Action Steps

Signaling and screening pop up in many real-life scenarios, including social situations, politics, and business. Answer these questions to further your understanding of these key concepts from game theory.

1. Think about certain groups you belong to in your life, like your family, employees of your company, or citizens of your country. What are some possible 'shibboleths' you could use to test whether another person is also a member of the same group?

2. Two friends are arguing over whether it makes sense to attend community college before transferring to an elite university after two years. One friend argues that it is cheaper and you end up with the same diploma, but the other friend argues you miss out on two years of an elite education by attending community college. From a signaling perspective, which argument is more convincing? Does our analysis change if we consider it undesirable to have community college on our resume?

3. Software developers seem to have a more casual dress code than finance workers or consultants. Consider our discussion about the importance of observable signals and unobservable skills. Why might the clothing expectations differ in each industry?

An Ounce Of Prevention...

In the beginning of the chapter, we described the story from the Hebrew Bible where the term 'shibboleth' derives from. While the Gileadites came up with a clever solution to solve the problem at hand, could better strategic thinking have avoided the need to come up with a solution in the first place?

If the Gileadites had some foresight, they might have realized that a victory would bring with it its own problems, due to the difficulty of distinguishing friendly soldiers from enemy soldiers. In thinking about this end state, the Gileadites could then adjust their behavior in the present to forestall such a situation.

For instance, the Gileadites could have had each soldier memorize a specific challenge phrase before the battle, which only their side would know about it. Then, they could rely on a very precise piece of knowledge to determine friend from foe, rather than a signal, that, while robust, may be imperfect.

Even when everything goes your way, strategic thinking can make your life just a bit easier. The above story serves as a simple example of the concept of backward induction, which is the next lesson from game theory that we will explore in depth.

Despite its intimidating name, this is likely a strategy you already put to work in your daily life. In the next chapter, we'll see how backward induction can serve to improve our lives both day-to-day and in the long term.

Key Takeaways

- To speed up our information-gathering process and allow us to navigate a complex world, we frequently rely on signals.
- Signals are a form of communication that indicates some underlying informative content. Signals are rarely direct forms of communication, but are ostensibly done for their own ends.
- Screening is a secondary act meant to ascertain how much information a signal actually contains. Not all signals are created equal, so we screen them to deduce what we should learn from them.
- Sometimes, refusing to show a common signal can provide more informative content than would otherwise be possible, an idea known as countersignaling.

6

TACTICAL RETREAT - HOW STEPPING BACKWARDS CAN MOVE YOU FORWARD

As the influential British economist John Maynard Keynes once remarked, the stock market resembles a strange beauty contest wherein the goal is picking the contestant that the majority of people would think is the most beautiful.[1] In the markets, it is not enough to merely pick a stock that you think is a wise investment. You need the rest of the market to agree with you in order to drive the price higher. Therefore, one's opinion is less important than one's estimation of others' opinions.

In 1997, Nobel-prize-winning economist Richard Thaler worked with the *Financial Times* newspaper to test out a version of Keynes' beauty contest. He asked the paper to collect answers to the following question:

Guess a number from 0 to 100 with the goal of making your guess as close as possible to two-thirds of the average guess of all those participating in the contest.[2]

Just like in the stock market, those who answer must consider how everyone else will answer if they want to be successful. Take a moment to consider how you might approach the question, and what answer you'd give.

A starting point of 50 might seem reasonable as the average guess for this contest, considering that would likely be the average guess made without the two-thirds clause. If we expect the average answer to be 50, then our guess should be 33, to get as close as possible to two-thirds of the average answer.

Of course, if other people follow this logic, then the average answer to the contest will actually be 33. Therefore, we can do even better by lowering our answer to 22, as it is two-thirds of 33. At this point, you might be able to see where this is going – if we expect the average answer to be 22, then we should guess 15, and so on and so forth until we fall all the way down to 0.

This iterative approach, wherein we take an optimal step at each fixed point to determine our final answer, is a popular problem-solving tool for navigating dynamic questions. By breaking down a complex whole into simple, one-step parts, many intimidating problems become much more approachable.

In this chapter, we'll explore one of the most important iterative approaches in game theory, a process known as backward induction. With backward induction, we start at some end state, and then work backward at each previous state to determine our optimal strategic decision.

We'll start by describing the theory behind backward induction, before seeing how the tool fares in real games.

Backward Induction

While backward induction might sound complex, the odds are that you already use the method in your daily life. For instance, when deciding what time to set your alarm for work, you might first consider what time you have to be at work, and then move backward in time based on how long each part of your routine will take. So, despite its theoretical underpinnings, the approach can be fairly intuitive with respect to certain day-to-day situations.

While backward induction can sometimes be used in one-shot games, it is most useful in multi-step or repeated games. When making iterative decisions, backward induction shines.

By way of example, backward induction can be used to analyze the finitely repeated prisoner's dilemma, a previous topic of discussion in chapter two. In the prisoner's dilemma, as we saw, the optimal strategic decision for both players is to selfishly defect. Defection is, therefore, a Nash equilibrium, since no player can improve their position based on the other's actions.

When we turn to the finitely repeated prisoner's dilemma, which is a multi-stage game, will our analysis change? At first glance, cooperation might seem more likely in the repeated game, since repeated interaction could build trust or offer the opportunity for punishment. Unfortunately, backward induction shows that even for the repeated prisoner's dilemma, defection is still theoretically optimal.

Consider that during the last stage, each participant will have a dominant strategy to defect, since there is nothing to gain by cooperating in the final round. Therefore, in the second to last round, each participant will also have a dominant strategy to defect – after all, if your opponent plans to defect in the *next* round anyway, they have no credible threat to punish you. Therefore, defection is optimal in the second to last round. This analysis can be carried back as far as necessary to show that each player has a dominant strategy to defect in each and every single round.

This approach showcases some key features that make backward induction so powerful. By turning the repeated prisoner's dilemma game into a series of iterative subgames, we can apply the standard tools of game theory to each stage without getting overwhelmed. Over the past several chapters, we've learned about concepts like dominant strategies and Nash equilibriums, all of which can be used through the framework of backward induction.

If you recall chapter two, however, you may remember that we described the 'tit for tat' strategy as one of the best strategies for the repeated prisoner's dilemma. Given our foregoing analysis, this might seem in direct contradiction with the strategic recommendations of backward induction. What can this discrepancy tell us about the utility of backward induction in practice?

Tit For Tat

The tit-for-tat approach, as described previously, is a strategy that can be deployed for a variety of games, most notably the repeated prisoner's dilemma. Tit for tat essentially calls for cooperation upfront, with punishment in case your competitor makes a selfish decision. This punishment is quickly withdrawn, however, as soon as the competitor begins cooperating again.

Tit for tat has several attractive features. First, it is an initially cooperative approach, which eschews selfishness in favor of the social gains of cooperation. Second, it imposes a cost on competitors who defect by defecting in the following round. Third, it is forgiving, allowing cooperation to rapidly resume rather than holding a self-defeating grudge.

Tit for tat is a notably effective approach in the iterated prisoner's dilemma. In the 1980s, University of Michigan political science professor Robert Axelrod held computer tournaments where various academics were invited to submit strategies to play the repeated prisoner's dilemma with each other. Remarkably, the ultrasimple tit-for-tat strategy, submitted by mathematical psychologist Anatol Rapoport, performed the best.[3]

As we mentioned, this seems to fly in the face of the conclusion reached by backward induction. It's important to remember, however, that the inherent tension in the prisoner's dilemma is that the socially optimal outcome is not necessarily the Nash equilibrium outcome. In other words, players would certainly do better if they both cooperate, but strict rationality advises them against cooperation at all.

If both players are willing to deviate from rationality slightly, however, they might discover a willingness on the other side to also cooperate for mutual benefit. This is why tit for tat is so powerful – it is not naïve enough to think that cooperation can be guaranteed, but it allows the possibility of doing so.

In terms of backward induction, we should always remember to take the tools of game theory with a grain of salt. While they can give us a place to start when it comes to strategic decision-making, human beings are not the rational figures game theory presumes us to be. In some cases, this is to our detriment. When it comes to social cooperation, however, it's a good thing we're not all Machiavelli. In fact, experimental evidence shows that most people in real games are willing to cooperate, despite what backward induction would advise.

Ultimatum Games

Ultimatum games are a popular subject of experimental economics, precisely for the way they deviate from the expectations of economics and game theory. In an ultimatum game, two participants are told about the same pot of money. One participant, the proposer, is tasked with proposing how to split the pot between the two participants. The second participant, the responder, can either accept or reject the proposer's offer, in which case neither participant gets any money.

Typically, the game is studied in a one-shot manner, meaning there is no repeated interaction. Backward induction shows that the responder should accept any positive amount offered to them. After all, even if the proposer allocates a tiny portion of the pot to the responder, some money is better than no money. Knowing this, the proposer should offer the responder the smallest amount of money they can, and keep the rest for themselves.

In practice, however, this is not what occurs. In fact, real-world experiments show that proposers offer an average of 40% of the pot to responders. Offers of less than 30% are often rejected.[4] In contrast to what economic theory proposes, many people feel that being offered a small portion of the pot is deeply unfair, and so choose to reject the offer in order to spite the proposer. People are willing to forgo monetary gains in order to avoid being disrespected, a fact evident to anyone who has seen a mistreated employee leave a high-paying job.

Knowing this experimental evidence, we can improve the way we apply backward induction to the game. While our original thought may have been that any positive offer would be accepted, we now see that we need to cross a minimum threshold to avoid offending the responder. The logic of backward induction remains intact, but we are able to more effectively utilize the tool with additional evidence about our environment.

Both these games show us that we should be wary of simply applying backward induction to the real world. As we have discussed repeatedly, humans are not the rational beings that game theory presumes them to be. In reality, while backward induction is a valuable tool, it should be used in conjunction with a search for evidence about people's behavior that might improve the way we use it. In addition, whether backward induction can lead to socially optimal outcomes ultimately depends on the specific game being played, as we'll see next.

Escalation Games

Escalation games are a broad class of games in game theory which refer to situations in which dynamic escalation between participants ratchets up tensions, likely leading to an all-out conflict. Backward induction can be used to analyze the extent to which any escalation game can be resolved peacefully, once the escalation begins.

For instance, consider two firms, A and B, competing for the market in a certain geographic area. Firm A cuts its prices in an attempt to capture a greater market share, which induces firm B to do so as well. In response, firm A drops its prices again, leading to a similar reaction from firm B.

From the outset, this dynamic may be difficult to analyze, requiring an understanding of the profit and loss impacts of each specific price change. Using backward induction, however, we can skip many layers of analysis to determine where the escalation game will lead.

Suppose the two firms have undergone many rounds of price competition, to the point where firm B has reached its marginal cost for producing each item. In other words, if firm B cuts prices any further, it will be selling for a loss. This induces firm B to exit the market, knowing that it cannot compete with firm A.

Of course, if firm B chooses to cut its losses at the final stage, it would do even better exiting the market at the first stage, saving the operating costs of fighting a losing battle. Therefore, if both firms carefully analyze the situation, combined with backward induction, firm B will choose not to compete with firm A on price. While this may or may not result in firm B leaving the market entirely, such analysis avoids a destructive sequence of escalation for both firms.

In the sense that mutual avoidance is preferred to mutual destruction, and that mutual destruction may yet occur, escalation games are closely aligned with chicken games, as introduced in [chapter one](). However, one clear difference is that chicken games tend to represent one single iteration of decision-making. In contrast, escalation games are repeated processes, representing multi-stage games. Experimental evidence indicates that the details of the bargaining process of these multi-stage games can affect the likelihood of poor outcomes.[5] Finding potential pathways to avoid such outcomes can therefore be a motivating factor to include backward induction in any analysis process of an escalation game.

Action Steps

Backward induction, while not flawless, is another tool from game theory that can improve our decision-making in the real world. Answer the following questions to further understand this analytical tool.

1. Consider a high school student who dreams of becoming a doctor. How could they use backward induction to plot a path toward achieving such a goal? What limitations might backward induction have in this process?

2 The propensity toward cooperation, rather than outright selfishness, seems inherent in human behavior. Given what we know about the success of tit for tat, why might this capacity for cooperation have been evolutionarily successful?

3. Suppose you are the responder in an ultimatum game, and you've been offered $5 from a $100 pot. Would you accept? Does your answer change if it is $500 from a $10,000 pot? Would your answer change depending on whether other people were observing your decision?

4. Despite the tensions of the Cold War, the United States and the Soviet Union did not initiate a nuclear war,

despite escalation. Would advising nuclear strategists to use backward induction have helped or harmed the situation?

Stepping Backward

As if we needed more evidence that game theory cannot be naively applied to the real world, the winning answer to the Financial Times contest we discussed in the beginning of the chapter was 13. While this answer reflects that most respondents did several rounds of the logical process we described, our theoretically correct answer of 0 would have won us no points.

In this book so far, we've discussed some different tools from game theory. Like backward induction, they can all be tremendously useful when applied properly. However, as the FT contest highlights, it would be very irrational to assume that everyone is rational. Therefore, all game theoretic tools should be used with humility about the uncertainties of the world around us.

In the next chapter, we'll review some of the initial developments of these tools by stepping back in time to the early history of game theory. While the field has advanced since the early days, many of the foundational concepts outlined by the first researchers continue to serve theorists and practitioners today.

Key Takeaways

- Backward induction is the iterative process of starting at the end state and working backward to determine an optimal strategy. It is applied to multi-stage or repeated games, since one-shot games do not involve iterated reasoning.
- Sometimes, the strategy advised by backward induction is empirically worse than others. For instance, the simple tit-for-tat strategy often results in better outcomes than the selfish strategy that backward induction calls for.
- In ultimatum games, backward induction would call for the proposer to share a small portion of the pot. In reality, though, this offer would likely be rejected. Therefore, backward induction needs to be modified to include realistic reactions from other participants.
- In escalation games, backward induction can be used to analyze situations that should be best avoided. For instance, a high-cost firm has no business getting into a destructive price war which they will likely lose.

7

A STRATEGIC LEGACY - STANDING ON THE SHOULDERS OF GIANTS

To describe John von Neumann, most peers, critics, and commentators rely on a single word: genius.

In today's age, the term is thrown about haphazardly. But von Neumann's fierce intellect, which merged theory and practice, makes him one of those rare historical figures who indisputably deserve the title.

In a (possibly apocryphal) story that highlights John's distinct mind, an interlocutor tries to stump him with the following question:

Two cyclists start 20 miles apart and head towards each other, each going at a speed of 10 miles per hour. At the same time, a fly traveling at a rate of 15 miles per hour starts at the wheel of one cyclist, flies to the wheel of the other, and repeats this process until he is crushed between the two wheels when the cyclists meet. What is the total distance the fly will cover?

Taken at face value, the question seems to require solving a complex infinite summation of the fly's trips between the two wheels. There is a neat trick, however, to solve the question instantly. Since the cyclists will take an hour to reach each other (each covering half of the 20-mile separation) and the fly travels at 15 miles each hour, the answer is 15 miles.

When the question was put to von Neumann, he quickly responded with the correct answer. Disappointed, the questioner told von Neumann that he must have already been aware of the trick. Irritated, von Neumann responded, "What trick? All I did was sum the infinite series."[1]

While von Neumann made significant contributions to physics, mathematics, computer science, statistics, and economics, he is of interest to us as the founder of game theory. As we mentioned in chapter two, von Neumann co-authored the influential 1944 book *Theory of Games and Economic Behavior*, which turned game theory from an eclectic mix of results into a systematized field.

In this chapter, we'll explore the foundations of game theory by reviewing the field's historical evolution. In doing so, we'll contextualize the knowledge we've reviewed so far, learning how the field's germination during the Cold War continues to influence its development today. While academics aside from von Neumann have made significant contributions to game theory, the life of the genius provides a fitting framework for understanding the story of the field as a whole.

John von Neumann

Born in Hungary in 1903, von Neumann's intellect was clear from an early age. When von Neumann was six, he was able to converse in ancient Greek. When he was 10, at the outbreak of World War I, von Neumann was able to engage in complicated discussions on military and political strategy with adults. By the time he got to university, he was acing his exams without attending classes. In a remarkably short span of just five years, von Neumann went from starting university to earning his Ph.D. in mathematics.

After some years in Germany studying mathematics, von Neumann accepted an appointment at Princeton University as a visiting professor. His routine of working in America during the academic year and returning to Europe during the summers was disrupted only by the rise of the Nazis.

In 1933, von Neumann joined the Institute for Advanced Study, an independent research outfit at Princeton. Notably, this position did not come with any teaching requirements, allowing him to spend nearly all of his time on research. Increasingly, von Neumann began to explore the mathematical structure of games and decision-making. In 1944, along with co-author Oskar Morgenstern, this would culminate in the publishing of *Theory of Games and Economic Behavior*, a book that would essentially launch game theory as a field of study.[2]

While it would be naïve to solely attribute von Neumann's interest in competitive decision-making to the intermittent instability of his early years or his lifelong passion for toys and games, both probably played a role. More likely, the increasingly dangerous geopolitical environment in the years building up to World War II helped von Neumann focus his efforts on theories that could guide the course of nations.

While this trend led to the aforementioned book, it also resulted in von Neumann lending his intellect to a number of American government projects. First and foremost was the Manhattan Project, where von Neumann helped create the atomic bomb.[3] In addition, von Neumann was a consultant for another organization that would leave its own independent mark on the historical path of game theory.

RAND

The RAND Corporation, which continues to operate today, was founded after World War II to support the United States Air Force through research and development (hence the name RAND). Eventually, the think tank would expand to inform American government policy in a variety of fields, including national defense, healthcare, and economic development.

Among the influential research stemming from RAND, the outfit's work in game theory has helped shape the development of the field. In chapter two, we discussed how the prisoner's dilemma was originally created by two RAND researchers, Merrill Flood and Melvin Dresher. As mentioned, John von Neumann also served as a consultant for RAND.

Somewhat facetiously, RAND offered von Neumann a monthly retainer fee for access to any ideas that came to him while shaving. According to a 1947 letter written to von Neumann from RAND reproduced in William Poundstone's book *Prisoner's Dilemma*:

> *We would send you all working papers and reports of RAND which we think would interest you, expecting you to react (with frown, hint, or suggestion) when you had a reaction. In this phase, the only part of your thinking time we'd like to bid for systematically is that which you spend shaving: we'd like you to pass on to us any ideas that come to you while so engaged.*[4]

Moreover, a later letter specifically requests von Neumann to help advance the corporation's work on game theory.

Of course, given that RAND was an organization designed to advance American strategic interests, the group did not merely want to pursue game theory research for the sake of its intellectual content. Game theory was leveraged by RAND specifically for thinking about the devastating consequences that would occur if the Cold War turned hot.

Would You Like To Play A Game?

Given the post-World War II backdrop that the RAND Corporation was founded in, the growing global tensions associated with the Cold War was a key area where the group contributed to American policy development. In particular, RAND played a tremendous role in forming American nuclear strategy. RAND's method of thinking about nuclear war was, in turn, heavily influenced by the precepts of game theory.

While previous generations of American policymakers may have been willing to view the conflict in terms of bilateral negotiations, RAND thought differently. Countries, just like competitors in game theory, would act ruthlessly and rationally in their own self-interest. Therefore, analyzing nuclear war could be done most effectively by thinking in terms of choices and payoffs in a clearly defined structure.

While thinking about nuclear war like a game may seem irresponsible, the strategy yielded insights that continue to guide American nuclear strategy. For instance, a 1954 RAND report advocated the development of second-strike capabilities, which could launch a nuclear counterattack in case a Soviet surprise attack devastated America. Nuclear submarines, capable of hiding at great depths and launching a strike on short notice, remain in use today as an effective second-strike deterrent.[5]

Note that this advice hews closely to guidance offered in chapter one on avoiding the worst outcomes in chicken games, which is to make a clear commitment to a devastating response if your enemy follows through on their attack. This is no coincidence, as RAND's advice draws on a similar game-theoretic structure. Second-strike policy eventually matured into what is known as 'mutually assured destruction,' a strategy initially promoted by RAND that has remained, in some form or another, America's nuclear policy since the end of World War II.

This cold analysis of the situation sometimes led game theorists to advocate for policies that might be considered inhumane, dangerous, or downright insane. For instance, as we mentioned in [chapter four](), John von Neumann was an advocate for a nuclear first strike against the Soviets. In his view, inspired by game theory, the second mover in a nuclear war only has a disadvantage, so the only rational move is to attack first. In a famous quote, von Neumann summarized his position: "If you say why not bomb them tomorrow, I say why not today? If you say today at 5 o'clock, I say why not 1 o'clock?"[6]

Thankfully, the advice of the genius was not put into practice, which might be why we still have a world to live in today. In any case, von Neumann was not the only thinker at RAND engaged with game theory. In fact, one of his colleagues arguably played an even more significant role in the development of the field as a whole.

John Nash

A brilliant mathematician whose life was marked by mental health troubles, John Nash is known for extending the work of the original founders of game theory by providing new analytical tools. Before starting at RAND, Nash studied at Princeton. Famously, his recommendation letter to the institution came to just a few lines:

> *Dear Professor Lefschetz,*
>
> *This is to recommend Mr. John F. Nash, Jr. who has applied for entrance to the graduate college at Princeton.*
>
> *Mr. Nash is nineteen years old and is graduating from Carnegie Tech in June. He is a mathematical genius.*
>
> *Your Sincerely, Richard J. Duffin"*[7]

Nash lived up to his reputation at Princeton. His Ph.D. thesis introduced a concept that would come to be known as a 'Nash equilibrium.'[8] As described in chapter two, a Nash equilibrium exists when no player has an incentive to change their strategy, given what their opponents have done. Nash's analysis helped expand the universe of games that game theory could successfully navigate, including non-zero-sum games and games with multiple competitors.

In addition to contributing one of the building blocks of modern game theory, Nash made significant contributions to pure mathematics, including differential geometry and partial differential equations. These achievements were not without difficulty and turmoil. Nash suffered from schizophrenia, which caused him hallucinations and paranoid delusions. Moreover, Nash lost his top-secret security clearance as a result of being charged with indecent exposure during a police sting operation. Although the charges were eventually dropped, Nash was fired from the RAND Corporation.

Despite these travails, Nash's impact was extraordinary. Shortly before his death, Nash was awarded the Abel Prize, one of the most prestigious awards in the field of mathematics. For his work in game theory, Nash was awarded the 1994 Nobel Prize in Economics, which he shared with two other prominent game theorists, Reinhard Selten and John Harsanyi.[7]

Reinhard Selten

Like so many other prominent game theorists, Reinhard Selten's life was not without difficulty. Selten was expelled from school at the age of 14 for being a Jewish German during the 1940s. After World War II, his family returned to Germany, where he would build his academic career. In Selten's case, this difficulty certainly influenced his decision to study economics. As he put it himself in his official Nobel biography:

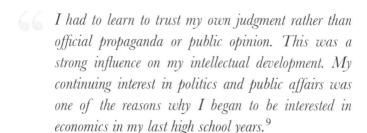

I had to learn to trust my own judgment rather than official propaganda or public opinion. This was a strong influence on my intellectual development. My continuing interest in politics and public affairs was one of the reasons why I began to be interested in economics in my last high school years.[9]

While Selten would win the Nobel for his work in game theory, he also made significant advances in experimental economics. Experimental economics is concerned with conducting laboratory experiments to test the validity of aspects of economic theory. As two fields that blend practical elements with theoretical knowledge, game theory and experimental economics go hand in hand.

Just as John Nash elaborated on the ideas of John von Neumann, Richard Selten was noted for expanding the concept of a Nash equilibrium. Selten developed the concept of a 'trembling hand' equilibrium.[10] In a trembling hand equilibrium, players assume that their opponents have a small chance of playing an imperfect strategy. This mirrors real life, where our competitors can sometimes make mistakes.

For instance, the Nash equilibrium of two rival nuclear powers might be that both sides continue to threaten the other with their weapons. If one side continues to threaten, the other must do so as well as an act of deterrence. But the existence of second-strike capabilities means that an outright attack is not an optimal strategy.

However, consider that a mistake by one side, like a rogue commander launching nukes, is still possible. For that reason, the trembling hand equilibrium ensures a series of fail-safes to prevent an accidental launch. The trembling hand equilibrium is therefore more robust to potential mistakes than the strict Nash equilibrium, which cannot account for fail-safe protocols.

Clearly, the Nobel committee had a penchant for work oriented toward real-world applicability, as the next economist also refined game theory for more practical use.

John Harsanyi

As we've seen, a life characterized by certain difficulties and instability may portend significant success in game theory. John Harsanyi is no exception. Like John von Neumann, Harsanyi was Hungarian-born and of Jewish descent. In 1944, Harsanyi's Jewish background almost landed him in a concentration camp, but he narrowly escaped from a railway station in Budapest. He took refuge in the home of a Jesuit priest and managed to survive the rest of the war.

In the post-war period, Harsanyi earned a Ph.D. in philosophy at the University of Budapest and became a junior faculty member. His outspoken anti-Marxist views, however, would force him to resign, given the political environment in Hungary. Harsanyi and his wife would make their way to Australia, where Harsanyi would earn an M.A. in economics. Finally, their winding journey would land them in the United States, where Harsanyi attained a second Ph.D., this time in economics.[11]

Harsanyi's game theory work particularly explored situations in which players have incomplete information about their competitor's likely actions. For the Nobel prize, Harsanyi was cited for papers that blended probability analysis with game theory.[12] In doing so, his work expanded the types of structures game theory could explore.

Previously, game theory could only analyze the types of games we've seen so far in this book with known payoffs and strategies. Harsanyi synthesized Bayesian analysis into game theory, meaning players had subjective beliefs about the likely strategies of their competitors. This closely mirrors the way real people operate in the world. While we never quite know what others will decide, we can certainly form a belief that helps motivate our own choices.

Action Steps

To engage with game theory's storied past, write your thoughts on the following questions.

1. Many of the figures discussed in this chapter had lives marked by difficulties. Do you think this is a coincidence or is there something in particular about these challenges that motivated these people to research game theory?

2. In the public consciousness, RAND is well-known for its Cold War nuclear policy. Do you think that RAND's contributions to nuclear strategy (including developing the mutually assured destruction doctrine) were a benefit or detriment to society?

3. Game theory is often criticized as overly abstract or rational and therefore difficult to apply to the real world. How do the advances discussed in this chapter mitigate this critique? Do they solve it entirely?

4. Without John von Neumann developing game theory during the emergence of the Cold War, how do you think the resulting conflict may have differed (if at all)? Can you think of any other examples of academic fields that grew quickly due to global trends?

Beyond Nuclear Strategy

While John von Neumann may be credited with founding game theory, John Nash, Reinhard Selten, and John Harsanyi all made significant contributions to forming the modern version of the field that continues to be an active focus of research. Although game theory grew up during the Cold War, it continues to find new applications in economics, political science, and, increasingly, biology, with the study of evolutionary systems.

The lasting legacy of game theory in the public mind remains the nuclear strategy used during the Cold War. The logic of game theory, which pits competing rational opponents against each other, describes the nuclear tensions of the period. Insights from research into the field helped motivate policies like second-strike deterrents and mutually assured destruction. Whether game theory made nuclear disaster more or less likely is ultimately a matter of debate, but there is no doubt the field had a significant impact on policy design during the period.

While it may have taken a genius to create game theory, we don't have to be geniuses to apply the field to the real world. In the next chapter, we will explore a variety of situations in which utilizing basic game theory can bring us real benefits. While modern advances in game theory can rely on complex mathematical theorems, sometimes, just slowing down and thinking strategically can make all the difference in the world.

Key Takeaways

- John von Neumann is credited with being the founder of game theory for an influential textbook he authored with Oskar Morgenstern in 1944. The book, *Theory of Games and Economic Behavior,* organized many of the results von Neumann had published previously.
- The RAND Corporation, a non-partisan think tank built to help advise American security policy in the post-World War II period, played an influential role in advancing the field of game theory.

- John Nash, at one point a consultant at RAND, invented the idea of a 'Nash equilibrium' in his PhD thesis. A Nash equilibrium, which describes a situation where no parties can improve their strategy given what their opponents have chosen, remains foundational in game theory.
- Reinhard Selten developed the idea of the 'trembling hand' equilibrium. This expanded the concept of a Nash equilibrium to optimize strategies when competitors can sometimes make mistakes.
- John Harsanyi contributed significant work to researching games with incomplete information, where competitors do not have full knowledge of their opponents' strategies.
- John Nash, Richard Selten, and John Harsanyi were jointly awarded the Nobel Prize in Economics in 1994.

8

PRACTICE MAKES PERMANENT - HOW TO NEVER FORGET WHAT YOU LEARN

Have you ever read a book filled with valuable knowledge that you know could improve your life, only to return to your old habits and routines several weeks later?

If so, don't blame yourself. Just reading information, no matter how it's presented, is usually inadequate for facilitating long-term retention. Without that retention, it's very difficult to integrate that information into your life.

Throughout this book, we've aimed to include a series of practice exercises at the conclusion of each chapter to facilitate understanding of the material. These exercises are inspired by the research undertaken into memory and understanding, which emphasize the importance of free recall testing when it comes to turning information into knowledge.[1]

In this chapter, we'll take these exercises to the next level. By exploring various key game theory concepts in the form of case studies with associated questions, you'll improve your understanding of the associated concepts and learn how to put them into practice in the real world.

Before diving into the case studies, we'll briefly review some of the most important ideas we've looked at so far.

Key Concepts

<u>Game Theory</u> – Game theory serves as the overarching framework of discussion for this book. Game theory is the art of using strategic thinking to make optimal decisions in a competitive environment.

<u>Prisoner's Dilemma</u> – The prisoner's dilemma is a popular game theory setup describing two accomplices who have been arrested for a crime. If they refuse to testify against the other, they will each receive short sentences. If one testifies against the other, the one who testifies will get no jail time, while the accused will receive a long sentence. If both prisoners testify, they will each receive a long jail sentence. Since both prisoners have a dominant incentive to testify, but the best outcome for everyone is not testifying, the example illustrates how self-interest can preclude a socially desirable outcome.

<u>Strategic Thinking</u> – Strategic thinking is the approach that underlies decision-making in game theory. It involves clearly defining goals, considering the potential strategies

of competitors, and weighing the possibilities of different actions before deciding on the best path forward.

Signaling And Screening – Signaling and screening is the two-step process by which we deduce information from our environment. Signals indicate the existence or non-existence of some underlying piece of information. Screening is the act of evaluating the veracity of a signal and judging its informative content.

Backward Induction – Backward induction is a popular problem-solving strategy in game theory that breaks down complicated situations into simpler elements. Backward induction involves determining a desired end state and then evaluating the necessary strategy at each previous step to achieve that end state.

Game Theory Case Study

Michelle is being considered as a candidate for a job. She is in the final stage of the hiring process where she engages in salary negotiation.

There are a number of competing dynamics in this scenario. Michelle is in the interview process with other companies but has no other solid offers. Therefore, she is worried about pushing too hard for a higher salary. At the same time, if the company rescinds the offer, she is still confident she will be able to line up another opportunity quickly.

The company, meanwhile, has a number of excellent candidates to pick from. While the hiring manager likes Michelle's personality and thinks she'd be a great fit for the team, they could be overruled by an executive if Michelle's salary requirements are too high. Still, the hiring manager is willing to engage in a good-faith negotiation to ensure Michelle feels properly valued.

The company has asked Michelle for her anticipated salary range. Michelle is aware that stating a tight range might be a detriment, as the company might believe there is little room for negotiation. At the same time, Michelle does not want to offer too expansive of a range, given that the company will likely make an initial offer at the lower end.

With all this in mind, Michelle has to figure out the exact salary range to offer to land the position while maximizing her salary.

<u>Case Questions</u>

1. Game theory emphasizes the importance of information when making decisions. What steps can Michelle take to learn about the various factors at play in this scenario?

2. Does this situation represent a one-shot game or a repeated game?

3. In some sense, this is a zero-sum game, since any salary gains that accrue to Michelle are at the company's

expense. In what ways, however, is this a non-zero-sum game?

4. If Michelle had another attractive job offer lined up, how might this change her strategic considerations?

Prisoner's Dilemma Case Study

You run a local business and need to decide on your advertising strategy for the year. You have just one main competitor in your town.

Currently, you don't spend any money on marketing or advertising. All your business comes through word of mouth. Since you haven't seen any advertisements for your competitor, you assume that they are in much the same position, with little money spent on marketing.

While your current strategy is certainly cheaper for you, you believe that allocating some money for advertising would help you gain a sizeable share of the market, especially considering that your competitor does not advertise at all. As it stands, both you and your competitor split the local market about equally.

You hired a consultant who projected that increased sales from an advertising campaign would dwarf any money spent on the campaign itself. You are smart enough to realize, though, that if you begin advertising, your competitor is likely to do so as well. The last position you'd want to be in is increasing your expenses and ending up with the exact same market split.

You also realize that your competitor is going through the same thought process. Theoretically, you could respond to a competitor's marketing campaign with one of your own. However, you worry that the long lead time it would take to fully organize a marketing campaign might result in your competitor entrenching a superior position. Dislodging this position might be uncertain, and require a much higher marketing budget.

With all these factors at play, you must decide whether or not you should launch an advertising campaign this year.

<u>Case Questions</u>

1. While this situation closely mirrors the prisoner's dilemma, there are some key differences. List some examples of the distinct features of this scenario.

2. Using the payoff structure detailed in <u>chapter two</u>, create a plausible payoff structure for this scenario that results in self-interest preventing the socially optimal outcome.

3. The prisoner's dilemma relies on each prisoner not being able to hold the other to account. Suppose you and your competitor knew each other. How might you design a punishment mechanism to ensure neither of you advertises, maintaining the high-profit status quo?

Strategic Thinking Case Study

Mike is a venture capitalist at a prestigious fund. He is considering a new round of investments and trying to decide which will be the most attractive.

Currently, AI investments are the hot trend in the venture capital world. Many of Mike's competitors and colleagues have been making new investments in the field. He is aware that the competition for investing in emerging AI companies has resulted in founders demanding extractive terms.

Overall, Mike is worried that AI might simply be another bubble. If the technology truly takes off and revolutionizes the industry, then such investments will pay off. In the scenario that excitement fades, however, he will face tremendous losses.

In comparison, Mike is reviewing many other investments that provide attractive return potential in established industries. While none of these investments is likely to pay off as a successful AI investment would, they are also less prone to severe failure.

While the explicit goals of Mike's venture capital fund are to provide an attractive return to investors, Mike also knows that only those who make the most successful investments will be promoted quickly within the fund.

Now, Mike has to decide which investments to pursue. Given the amount of capital allocated to him, he can only invest in a few companies.

Case Questions

1. Mike seems to have several different competing goals. If he were to sit down and clearly define his desired end state, what might the different options look like?

2. Who would you consider Mike's competitors in this scenario? Is it just venture capitalists at other funds, or his colleagues as well? What are the distinctive competitive features of each group?

3. What would you consider Mike's various potential strategies to be? How might he be able to define a new, blended strategy which incorporates strengths from different options?

Signaling And Screening Case Study

You need to buy a refrigerator for your new home but you're uncertain about which brand is best to buy. This is a very important purchase for you since you'll likely use the same fridge for many years.

While customer reviews might be a good place to start, you're worried that the information provided may not be accurate. For instance, the quality of the product may have dropped since the reviews were written. Additionally, you are aware that some firms will pay for good customer reviews.

You decide that your best strategy is to find a company that offers an excellent return and warranty policy for their fridges. You believe that if a company is willing to honor a strong return policy, they must believe in the quality of the products they are selling.

At the same time, you recognize the hassle of returning a fridge, requiring scheduling a pickup process, and being without the appliance for a few days. This might make actually leveraging the return policy a difficult endeavor. If enough customers act this way, the return policy won't say very much about the quality of the product at all, since the firm knows it will rarely be used.

Now, you have to decide which company to buy a fridge from. While cost may be a factor, you are much more concerned with purchasing a high-quality appliance.

Case Questions

1. Describe two signals in this case study and how the screening process for each differed.

2. Recognizing the difficulty a customer might have returning a refrigerator, how might an appliance firm strengthen the signal of their warranty?

3. What other possible signals could you use to evaluate the quality of potential options?

4. Using this specific case, describe how the dynamics reveal the necessity of relying on signals in everyday life. You might consider the training and travel necessary to independently verify the quality of a given fridge.

Backward Induction Case Study

Currently, Rebecca runs a software firm. While she has had plenty of success with the business, she is ready to hand the reins off to someone else and enjoy her retirement. She is unsure, however, of the specific route she'd like to follow.

First, Rebecca has a son who could potentially take over the firm. While the son has never shown much interest in software, Rebecca believes she could train him over the course of the year to manage the business well. This would have the benefit of keeping the company within the family.

Second, Rebecca could sell to a rival. Since the rival has competed with Rebecca's firm for years, they know her business very well. This would make a sale a fairly quick endeavor, although Rebecca does not expect the rival to pay top dollar. Additionally, Rebecca would have to swallow her pride in selling to a company she has competed with for years.

Finally, Rebecca could sell to a private equity group. A private equity fund would pay the most for her firm, but it would be a long, drawn-out process, which could ultimately result in the deal collapsing. Lawyers and accountants would need an extended period to analyze the firm's books and vet its operations.

All these competing options seem potentially attractive. While Rebecca certainly cares about the money she'll receive from selling the business, she also wants to preserve her reputation and legacy. Going down any one of the routes will require her to navigate different situations. Now, it's up to her to make a decision.

Case Questions

1. How might backward induction be a useful tool in this case? Even if Rebecca is still torn between the three potential end states, might the tool still have benefits?

2. Rebecca has a fairly good idea of what each path will require to complete successfully, but she can never be exactly sure. What does this fact reveal about the limitations of backward induction in the real world?

3. For at least two of the stages, a negotiation process will likely be required to determine the final sale price. How might Rebecca utilize backward induction in this specific scenario?

Action Steps

While we've answered a lot of questions in this chapter, let's review a few more to integrate our knowledge fully.

1. How would you describe the difference between game theory and strategic thinking? Can we perform strategic thinking outside the realm of game theory?

2. How can knowledge of the prisoner's dilemma improve outcomes for us in potentially contentious situations in

our personal or professional lives? Is merely being aware of the prisoner's dilemma enough?

3. Although the case studies we reviewed in this chapter differed significantly, many had common elements stemming from their roots in game theory. What factors were common to all (or most) of the situations we reviewed?

Practice Makes Perfect

Although reading the same content again and again can lead to memorization, actually putting knowledge into practice requires engaging with challenging questions designed to test how well you've integrated the information. With this chapter's series of case studies, as well as the practice problems presented at the end of each chapter, you can work to synthesize the knowledge presented in this book to improve your decision-making in the real world.

In the next chapter, we will follow up with this aim by presenting some of the key life lessons we can learn from game theory.

Key Takeaways

- While game theory is the art of optimizing decision-making in competitive environments, we can put game theory into practice by using strategic thinking to guide our choices.

- The prisoner's dilemma is applicable to almost any situation in which selfishness prevents socially desirable outcomes from occurring. This makes it a powerful framework for analyzing many distinct situations in life.
- Signals are useful because they can save us a lot of time searching for information ourselves. However, every signal must be appropriately screened to determine its informative content.
- While backward induction is not flawless, it provides an excellent strategy to review how individuals can achieve specific desired end states.

9

MAKING A PLAYBOOK - WHY HUMANS AND ROBOTS ARE DIFFERENT

Both economics and mathematics can be somewhat arcane fields, which can make real-world applications difficult. Sometimes, even the creators of advanced theories have trouble practically utilizing them.

Economist Harry Markowitz won the 1990 Nobel Prize for his work on how investors can optimally diversify their portfolios. His theory advised people to balance their investments in a specific way based on each asset's volatility and correlation with other holdings. The resulting work proved extremely influential in finance, eventually becoming known as "Modern Portfolio Theory."

However, in an interview that Markowitz gave in 1997, the economist revealed that even he did not invest the way his theory advised. As Markowitz recalled in the interview when setting up his allocations,

> *I should have computed the historical co-variances of the asset classes and drawn an efficient frontier. Instead, I visualized my grief if the stock market went way up and I wasn't in it–or if it went way down and I was completely in it. My intention was to minimize my future regret. So I split my contributions 50/50 between bonds and equities.*[1]

This is a far cry from leveraging the esoteric mathematics his original paper called for. Moreover, this might cause us to worry about applying our game theory knowledge to the real world. Game theory is perhaps the quintessential mix of mathematics and economics. If even Harry Markowitz struggled to put a similar blend into practice, is there hope for us?

Thankfully, the answer is yes. In fact, Markowitz's comments reveal the most important lesson that we'll have to keep in mind to effectively translate game theory to practice: Remember the human.

Regardless of any theoretical knowledge, a human being will ultimately make decisions. Knowing how subject we all are to our hopes and fears, ensuring that the lessons we draw from the field speak to our human experience is imperative.

Throughout this book, we've worked to translate the theories described into real-world advice. In this chapter, we'll continue to do so by presenting the most meaningful life lessons that can be drawn from game theory. As always, our goal is not to be impressed with theory for theory's sake, but to leverage such knowledge to lead happier, healthier, and wealthier lives.

People Are Self-Interested, But Not Absolutely

One of the most important lessons that game theory can impart is to be aware of people's inherent self-interest. At its core, game theory assumes that people are selfish. This is indicative of the way optimal strategies for a competitor only incorporate the payoff that competitor will receive. Even in situations where game theory models altruism, the field assumes that such altruism only occurs because the selfless person derives some emotional benefit from their sacrifice.

In reality, of course, people are not as selfish as game theory would make it seem. Throughout this book, we've seen numerous examples of the ways that people deviate from traditional models of self-interested behavior. In particular, the research experiments we've reviewed show how people are willing to shift from a purely selfish strategy, in contrast to what game theory would predict.

In chapter two, for instance, we saw that experiments that mirror the prisoner's dilemma sometimes induce cooperation instead of defection, contrary to what theory would predict. Of course, since cooperation can benefit an individual, this can be viewed as a selfish action regardless. Similarly, as discussed in chapter six, people tend to offer their opponent a sizeable portion of the pot in ultimatum games, but this action is likely motivated by fear of a low offer being rejected out of spite.

In sum, while people's motivations are more complicated than the purely selfish model imparted by game theory, it is vital to be aware of how others will defend their self-interest. This does not mean we should take game theory as a license to be greedy. But it does mean we should avoid being naïve and be willing to stick up for ourselves when others try to benefit at our expense.

Be Willing To Take Risks And Make Mistakes

Another game theory lesson that can help grow our share of the pie is to embrace taking small risks and making survivable mistakes. Although we should not actively try to mess up a task or disrupt our current circumstances, being willing to bear potential downside should ultimately improve our lives.

This is not an obvious lesson to draw from game theory. The field is often fixated on optimizing analysis using complex mathematics to *avoid* making mistakes. But the key insight is to realize that, for most people, spending time modeling and solving these equations is either impossible or impractical.

First, doing so requires fully understanding the situation we're analyzing, which is often impossible in real life. Second, academic research is a full-time job, so trying to conduct literature reviews and apply existing methods to novel situations is not feasible for anyone with an already-busy schedule.

Instead, we can focus on achieving optimal outcomes through trial and error. By embracing an iterative approach, we can make up for any lack of knowledge or mathematical prowess with real, practical experience. This approach mirrors the evolutionary process, where even individually unintelligent animals can become optimized for their current environment by culling missteps and hanging onto successes.

Despite how failure is often denigrated in society, research has helped demonstrate how those who have failed previously are arguably *more* fit to succeed in the future.[2] Since people's qualities are not fixed, their failures may simply result from a lack of experience or knowledge rather than due to their specific personal shortcomings.

Keeping this attitude in mind can help us embrace making small bets, even though they may not work out. As always, our goal should be to find our optimal strategy, whether that's through complex mathematics or a simpler trial-and-error approach.

Step Outside Your Own Perspective

Empathy remains one of the most vital life lessons to remember from game theory. In many ways, the field is simply the art of viewing things from another person's perspective. As discussed in [chapter four](), the inherent competitiveness of game theory motivates empathy. It is impossible to strategically outmaneuver your opponent without considering their likely actions. The only way to consider their actions, of course, is to understand how they would view the situation.

Empathy provides a dual benefit in the context of strategic decision-making. First, it certainly helps improve outcomes for ourselves. Appreciating your competitor's perspective can inform our own decisions. Second, empathy can lead to more socially optimal outcomes. In particular, empathy can help us find previously unknown avenues for strategic collaboration. As we've seen, individual benefits and social benefits in game theory have a complicated relationship. While they are not always aligned, empathizing can certainly help us identify situations in which they are.

In addition, research indicates that empathy triggers an altruistic response in people, which induces them to help others. In one study, people showed a propensity to help others in need, even when escaping from the situation was easy.[3] This indicates that kindness motivates prosocial behavior rather than simply the motivation to stop experiencing negative emotions as a result of empathizing with those in trouble. This provides a strong argument that tapping into our empathetic side can lead to better outcomes for everyone. Of course, empathy does not preclude recognizing how important competing with others can be to achieving our best results.

Embrace Healthy Competition

The importance of competition is baked into the foundations of game theory. Nearly every game analyzed is structured as an interaction between two competing parties. Without opponents with diverging interests, of course, there would be very little strategic decision-making to consider in our choices.

Game theory can remind us of the importance of embracing competition. Competing with others helps us become the best version of ourselves. This is seen very clearly on the sports field, where athletes are motivated to play their best in order to win the game. Such a lesson carries over into the classroom and the business environment, where competition continues to push people to stand out from the group.

Of course, competition can occasionally become antithetical to good performance. If competition is too high pressure, for instance, people may burn out. In addition, ruthless competitive structures can disrupt the benefits of cooperation as people start to view each other as enemies. For this reason, focusing on healthy competition, which is structured to ensure that people maintain space to decompress from a rivalrous atmosphere, is vital.

Embracing competition can be scary because we always risk falling behind and, to be blunt, failing. But this ties closely into the importance of accepting risk, a previous game theory lesson we discussed. Only by accepting the possibility of failure can we hold on to the possibility of success.

Of course, not every game is worth competing in. The next lesson of game theory involves knowing which games are the right ones to play.

Good Games And Bad Games

In life, not all games are created equal. Some games are advantageous, where we have a real possibility of achieving a goal, without risking undue consequences. Other games are risky, which have an overwhelming chance of leading us to ruin.

The distinction is captured memorably in *The Gambler*, a country song by American artist Kenny Rogers. In the song, a poker player reminds us, to know when to hold or when to fold. In poker, as in life, not all hands are worth playing. Sometimes, it's best to avoid playing certain types of games.

Game theory teaches us this life lesson through the way in which certain game structures inevitably lead to terrible outcomes. For instance, recall the chicken game we described in <u>chapter one</u>. While it's possible to avoid mutual destruction by cultivating a strong threat, repeated escalation is likely to lead to disaster. In the next chapter, we described the archetypal no-win game, the prisoner's dilemma.

In real life, these games might manifest as a cutthroat work environment, an escalating argument with your partner, or a research group prone to infighting and credit-stealing. Whatever the manifestation of the game, recognizing when certain situations will lead to poor outcomes, no matter how well you think you'll navigate it, is a vital skill. Sometimes, knowing which games to avoid is just as important as knowing how to play games well.

Visualize The Future To Determine The Present

Playing games well, of course, requires the type of careful strategic thinking the game theory calls for. The next life lesson from game theory is on the importance of planning our future paths to achieve the goals we want. This mental attitude shifts us from a mere desire to achieve our goals to taking practical steps to truly accomplish them.

It is common to focus on the confidence-boosting benefits of mental rehearsals and visualization when it comes to future challenges. Of course, the benefits of believing in oneself should not be understated. But research indicates that visualizing exercises can improve outcomes by helping us to develop a realistic plan to navigate daunting tasks. As one paper memorably put it, shifting from an "I can do it" mental approach to a "How can I do it?" one appears most effective for goal attainment.[4]

This ties neatly into game theory. Practicing good strategy requires understanding your present situation, clearly defining your goals, and analyzing which tactics can help you get there. This is especially true with the multi-step games that characterize the real world, where a series of iterative choices can be necessary.

By precisely visualizing our strategic choices, we can break down seemingly daunting goals into smaller tasks. For instance, if you are entering college with the aim of becoming a doctor, you first have to select an appropriate major. To successfully complete the major, you'll have to do well on your exams. To do well on your exams, you will have to attend classes and create a study schedule. By working through the process in our mind's eye, we can take actionable steps each day to achieve our desired outcome, as inspired by the strategic mindset of game theory.

Selfishness Can Inhibit The Common Good

The final major lesson of game theory that we will explore is the way in which selfish behavior can frequently disrupt socially optimal outcomes. In our first lesson, we noted that while people are not as selfish as game theory would predict, it's important to keep in mind the way other people's self-interest can disrupt our own goals. The current lesson is the social analog to this notion, which recognizes that other people's self-interest can also disrupt the goals of a community as a whole.

The prisoner's dilemma is the clearest game theory example of how self-interest can ensure a social optimum is not reached, as each prisoner has an incentive to testify even though both keeping silent would minimize the total jail time. The issue is also frequent in the real world, however. For instance, consider a crowd escaping a fire in a crowded movie theater. If everyone were willing to walk to the exits, evacuation could take place safely and calmly. But since each person has a selfish interest in sprinting to the exit if everyone else walks, the more likely outcome is a dangerous surge toward the door.

At a global level, this same dynamic results in the 'tragedy of the commons.' The tragedy of the commons describes a situation in which people do not have an incentive to take proper care of a shared resource.

Consider a dozen fishermen who all fish from the same lake. Suppose none of the fishermen want the lake to be overfished, lest the area run out of fish. Despite this desire, each fisherman has an incentive to overfish. Since the actions of one fisherman are unlikely to meaningfully impact the lake's fish stock, they will get to catch extra fish while not ruining the lake. Since each fisherman makes this same calculation, overfishing is very likely to occur.

In the world today, this dynamic plays a significant role in issues related to climate change, as explored in the introduction to chapter two. While ample research has been conducted on the best ways to minimize tragedy of the commons issues, many of these solutions have yet to be put into practice.[5]

In any case, leaders and managers should keep this lesson in mind when structuring their organizations. If incentives are not aligned properly, self-interest can easily overwhelm the common good.

Action Steps

In this chapter, we detailed some of the practical life lessons that game theory can teach us as we navigate the world. To better understand these lessons, engage with the following questions.

1. If some researchers and academics fail to take their own advice, why should we pay attention to it? What do you think creates the distinction between theoretical work that is useful and work that is not?

2. If taking risks necessarily opens up the door to failure, how can we practice effective risk management to ensure we do not reach a point of ruin?

3. Competition can push us to achieve great things but it can also create unnecessary stress and lead to burnout. How can we balance competition effectively to benefit from it without creating significant detriment?

4. It's not always easy to recognize which games are worth playing from an outside perspective. If a game is not worth playing, what does that mean to you? Does it have more to do with the structure of the payoffs or with the competitors you'll be playing with? How can we effectively recognize such games from the start?

A Playbook For The Game Of Life

Although game theory's mathematical and economic foundations can be intimidating, when the field is translated into human language, its practical use becomes evident. In this chapter, we've worked to synthesize the knowledge explored so far into a series of practical life lessons. While everyone will have to engage in some reflection to understand how these lessons best fit into their own unique circumstances, they provide some powerful tools for building a better life through better decision-making.

As this chapter serves as the penultimate one, these life lessons help synthesize the vast majority of the information contained in this book. Readers should be aware, though, that we have only scratched the surface of some of the fascinating and complex work taking place in modern game theory. Staying abreast of the latest developments in the field is an excellent way to continue to improve your decision-making in search of a happier, healthier, and wealthier life.

Key Takeaways

- While people are selfish, they are not the ruthlessly self-interested caricatures that game theory makes them out to be. With that in mind, it's important to remember to stick up for our own interests, just as other people do.

Selfishness can also preclude the best outcomes for society.
- Being willing to take small risks can be a powerful path to improving your circumstances. This trial-and-error approach can replace the need for overwhelming intelligence.
- Practicing empathy can, of course, lead to more socially optimal outcomes. It can also be used, however, to improve our own outcomes by helping us navigate our competitor's likely strategies.
- Embracing healthy competition can help us become the best version of ourselves. However, it is important to avoid competition that could lead to burnout or unhappiness.
- Not all games are worth playing, and not all situations are worth staying in. It's important to recognize the scenarios where not engaging is the best option.
- While visualizing your future goals can help make you more confident that you will achieve them, the more important factor may be the way in which visualization helps you construct an actionable plan to achieve said goals.

AFTERWORD

The 2014 film *The Imitation Game* depicts Alan Turing and a small group of cryptographers trying to break the Nazi's enigma code during World War II. Eventually, the group succeeds, giving them access to nearly all the Nazi's strategic communications, including the locations of troop movements and ships.

In one powerful scene, the group realizes that, even after cracking enigma, they must be exceedingly careful about how they use their knowledge. If the Nazis realized that enigma had been cracked, they would stop using it, and all the work would have been for nothing. At one point, this even involves letting a Nazi attack on civilian boats occur.

While the film certainly takes some dramatic license, this episode does mirror real considerations the Allies had to make when putting their knowledge of enigma into practice. The fact that enigma had been cracked was known only to the very highest levels of the British government. Any time the Allies acted on deciphered intelligence, they had to create a plausible cover story of where they got the information.

In many ways, this episode perfectly captures the essence of game theory. Success in life does not just depend on having access to the best information or the most resources. It also depends on using the information and resources you have strategically by carefully considering how other people will react. Breaking enigma required intelligence, but using that fact to win the war required strategy.

In this book, we've explored a broad overview of the facts and function of game theory. We went from discussing some basic game theory structures to seeing how the field operates in the real world. Then, we explored some of the most important manifestations of game theory in day-to-day life, including signaling and backward induction. Finally, we took a step back, examining game theory's origins, practicing our new knowledge, and summarizing some key life lessons from the field.

But the real work has only just begun.

From Knowledge To Action

While this book can provide you with the tools to start incorporating game theory into your life, actually doing so rests on your shoulders. Knowledge is no substitute for action.

As you go throughout your day and make decisions, actively consider how strategic thinking can be incorporated to improve outcomes. Work backward from desired solutions to generate a plan. Make sure you're giving off the right signals. Search for potential cooperative solutions, while avoiding mutually destructive games. Take small risks and embrace healthy competition to become the best version of yourself. And most importantly, empathize with other people and consider their perspective.

Success looks different for everyone. While it's unlikely you'll be using strategic thinking to keep a monumental code-breaking feat secret, your goals might include securing a raise, finding a partner, getting your dream job, starting a business, raising a family, or some mix of all of these.

Whatever your destination, game theory can help you get there. But only if you're willing to translate the knowledge you have into effective action. Good luck.

OVER 10,000 PEOPLE HAVE ALREADY SUBSCRIBED. DID YOU TAKE YOUR CHANCE YET?

In general, around 50% of the people who start reading do not finish a book. You are the exception, and we are happy you took the time.

To honor this, we invite you to join our exclusive Wisdom University newsletter. You cannot find this subscription link anywhere else on the web but in our books!

Upon signing up, you'll receive two of our most popular bestselling books, highly acclaimed by readers like yourself. We sell copies of these books daily, but you will receive them as a gift. Additionally, you'll gain access to two transformative short sheets and enjoy complimentary access to all our upcoming e-books, completely free of charge!

This offer and our newsletter are free; you can unsubscribe anytime.

Here's everything you get:

✓ How To Train Your Thinking eBook ($9.99 Value)
✓ The Art Of Game Theory eBook ($9.99 Value)
✓ Break Your Thinking Patterns Sheet ($4.99 Value)
✓ Flex Your Wisdom Muscle Sheet ($4.99 Value)
✓ All our upcoming eBooks ($199.80* Value)

Total Value: $229.76

Go to wisdom-university.net for the offer!

(Or simply scan the code with your camera)

*If you download 20 of our books for free, this would equal a value of 199.80$

THE PEOPLE BEHIND WISDOM UNIVERSITY

Michael Meisner, Founder and CEO

When Michael ventured into publishing books on Amazon, he discovered that his favorite topics - the intricacies of the human mind and behavior - were often tackled in a way that's too complex and unengaging. Thus, he dedicated himself to making his ideal a reality: books that effortlessly inform, entertain, and resonate with readers' everyday experiences, enabling them to enact enduring positive changes in their lives.

Together with like-minded people, this ideal became his passion and profession. Michael is in charge of steering the strategic direction and brand orientation of Wisdom University, as he continues to improve and extend his business.

Claire M. Umali, Publishing Manager
Collaborative work lies at the heart of crafting books, and keeping everyone on the same page is an essential task. Claire oversees all the stages of this collaboration, from researching to outlining and from writing to editing. In her free time, she writes online reviews and likes to bother her cats.

Lorey L. de Guzman-Dadula, Co-Publishing Manager

Lorey is a dedicated writer who shares her home with three cats, one dog, and her beloved husband. In her role as Wisdom University's co-publishing manager, Lorey taps into her prolific professional experience in management and education to effectively support a diverse array of amazing talents at every stage of the publication.

Brian Flaherty, Writer

Brian is a writer who focuses on exploring and explaining topics in business, finance, and economics. He has a background in the financial industry, having served as the Chief Strategist of a wealth management firm. Brian holds a B.A. in Economics from the University of Virginia.

Jevette Brown, Content Editor

Jevette is an editor with a background in copy editing, academic writing, and journalism. With bachelor's degrees in Communications and English and a graduate degree in Legal Studies, Jevette's varied experience provides a valuable perspective for diverse content and audiences. Her priority is helping writers fine-tune their material to display their unique voices best.

Sandra Agarrat, Language Editor

Sandra Wall Agarrat is an experienced freelance academic editor/proofreader, writer, and researcher. Sandra holds graduate degrees in Public Policy and International Relations. Her portfolio of projects includes books, dissertations, theses, scholarly articles, and grant proposals.

Danielle Contessa Tantuico, Researcher

Danielle conducts comprehensive research and develops outlines that are the backbone of Wisdom University's books. She finds pleasure in this role as it allows her to immerse herself in self-improvement topics. An avid reader and a songwriter, Danielle channels her passion for artistic endeavors and personal growth into everything she creates for Wisdom University.

Ralph Escarda, Layout Designer

Ralph's love for books prevails in his artistic preoccupations. He is an avid reader of non-fictional books and an advocate of self-improvement through education. He dedicates his spare time to doing portraits and sports.

Jemarie Gumban, Publishing Assistant

Jemarie is in charge of thoroughly examining and evaluating the profiles and potential of the many aspiring writers for Wisdom University. With an academic background in Applied Linguistics and a meaningful experience as an industrial worker, she approaches her work with a discerning eye and fresh outlook. Guided by her unique perspective, Jemarie derives fulfillment from turning a writer's desire to create motivational literature into tangible reality.

Evangeline Obiedo, Publishing Assistant

Evangeline diligently supports our books' journey, from the writing stage to connecting with our readers. Her commitment to detail permeates her work, encompassing tasks such as initiating profile evaluations and ensuring seamless delivery of our newsletters. Her love for learning extends into the real world - she loves traveling and experiencing new places and cultures.

REFERENCES

Introduction

1. Hollenhorst, M. (2020). *CEO Reed Hastings on how Netflix beat Blockbuster*. Marketplace. https://www.marketplace.org/2020/09/08/ceo-reed-hastings-on-how-netflix-beat-blockbuster/

1. Everyday Game Theory - Why Smart People Do Foolish Things

1. Doyle, P. (2017, February 20). The day in 1982 when the world wept for Algeria. *The Guardian*. https://www.theguardian.com/football/2010/jun/13/1982-world-cup-algeria

2. The Dilemma Of Trust - How To Avoid Jail Time

1. Poundstone, W. (1992). *Prisoner's dilemma: John von Neumann, game theory, and the puzzle of the bomb*. Anchor Books.
2. Laertius, D. (2014). *The lives and opinions of eminent philosophers* (Yonge, C.D., Trans.). CreateSpace Independent Publishing Platform. (Original work published 1853b)
3. Walker, P. & Walker, A. (2018). *The golden rule revisited*. Philosophy Now. https://philosophynow.org/issues/125/The_Golden_Rule_Revisited
4. Ngo, M. (2023, May 1). *A timeline of how the banking crisis has unfolded*. The New York Times. https://www.nytimes.com/2023/05/01/business/banking-crisis-failure-timeline.html
5. Ertan, A., Page, T., & Putterman, L. (2009). Who to punish? Individual decisions and majority rule in mitigating the free rider problem. *European Economic Review, 53*(5), 495–511. https://doi.org/10.1016/j.euroecorev.2008.09.007

6. Marwell, G., & Ames, R. (1981). Economists free ride, does anyone else?: Experiments on the provision of public goods, IV. *Journal of Public Economics*, *15*(3), 295–310. https://doi.org/10.1016/0047-2727(81)90013-x

3. The Hidden Chessboard - How Game Theory Shapes Our World

1. Isaacson, W. (2011). *Steve Jobs*. Simon and Schuster.
2. Taleb, N. N. (2007). *The black swan: The impact of the highly improbable*. Allen Lane.
3. Connor, J. M. (1997). The global Lysine price-fixing conspiracy of 1992-1995. *Applied Economic Perspectives and Policy*, 19(2), 412. https://doi.org/10.2307/1349749
4. Andreoli-Versbach, P., & Franck, J. (2015). Endogenous price commitment, sticky and leadership pricing: Evidence from the Italian petrol market. *International Journal of Industrial Organization*, *40*, 32–48. https://doi.org/10.1016/j.ijindorg.2015.02.006
5. Steinberg, R., & Zangwill, W. I. (1983). The Prevalence of Braess' Paradox. *Transportation Science*, *17*(3), 301–318. https://doi.org/10.1287/trsc.17.3.301
6. The Economist. (2020, June 11). *Driverless cars show the limits of today's AI*. https://www.economist.com/technology-quarterly/2020/06/11/driverless-cars-show-the-limits-of-todays-ai
7. Dixit, A. K., & Nalebuff, B. (2008). *The art of strategy: A game theorist's guide to success in business & life*. W. W. Norton & Company.

4. Making Good Choices - What You Can Learn From The Kindergarten Playground

1. Iyengar, S. S., & Lepper, M. R. (2000). When choice is demotivating: Can one desire too much of a good thing? *Journal of Personality and Social Psychology*, *79*(6), 995–1006. https://doi.org/10.1037/0022-3514.79.6.995
2. Camerer, C. F., & Weber, M. (1992). Recent developments in modeling preferences: Uncertainty and ambiguity. *Journal of Risk and Uncertainty*, *5*(4), 325–370. https://doi.org/10.1007/bf00122575

3. Field, A. J. (2014). Schelling, von Neumann, and the event that didn't occur. *Games*, *5*(1), 53–89. https://doi.org/10.3390/g5010053

5. Sign Language - When To Fit In And When To Stand Out

1. [1] Kemmer, S. (n.d.). *The story of the Shibboleth*. Words in English. http://www.ruf.rice.edu/~kemmer/Words/shibboleth
2. Ambrose, S. E. (1994). *D-Day: June 6, 1944: The climactic battle of World War II*. Simon & Schuster.
3. Hungerford, T. L., & Solon, G. (1987). Sheepskin Effects in the Returns to Education. *The Review of Economics and Statistics*, *69*(1), 175-177. https://doi.org/10.2307/1937919
4. Caplan, B. (2018). *The Case against education*. Princeton University Press. https://doi.org/10.23943/9781400889327
5. Rohwer, S. (1977). Status Signaling in Harris Sparrows: Some Experiments in Deception. *Behaviour*, *61*(1–2), 107–129. https://doi.org/10.1163/156853977x00504
6. Taleb, N. N. (2018). *Skin in the game: Hidden asymmetries in daily life*. Random House.
7. ABC News. (2020, February 8). *WATCH LIVE: Democratic presidential candidates debate in New Hampshire | ABC News Live* [Video]. YouTube. https://www.youtube.com/watch?v=_JCTY6MxJ4I

6. Tactical Retreat - How Stepping Backwards Can Move You Forward

1. Keynes, J. M. (2018). *The general theory of employment, interest, and money*. Palgrave Macmillan.
2. Thaler, R. H. (2016). *Misbehaving: The making of Behavioral Economics*. W. W. Norton & Company.
3. Axelrod, R. (1980). Effective Choice in the Prisoner's Dilemma. *Journal of Conflict Resolution*, *24*(1), 3–25. https://doi.org/10.1177/002200278002400101
4. Oosterbeek, H., Sloof, R., & Van De Kuilen, G. (2004). Cultural Differences in Ultimatum Game Experiments: Evidence from a Meta-Analysis. *Experimental Economics*, *7*, 171–188. https://doi.org/10.1023/b:exec.0000026978.14316.7

5. Chuah, S., Hoffmann, R. S., & Larner, J. (2013). Elicitation effects in a multi-stage bargaining experiment. *Experimental Economics*, *17*(2), 335–345. https://doi.org/10.1007/s10683-013-9370-z

7. A Strategic Legacy - Standing On The Shoulders Of Giants

1. Halmos, P. R. (1973). The legend of John von Neumann. *American Mathematical Monthly*, *80*(4), 382-394. https://doi.org/10.2307/2319080
2. Poundstone, W. (1992). *Prisoner's Dilemma: John von Neumann, game theory, and the puzzle of the bomb*. Anchor Books.
   See also: Von Neumann, J., & Morgenstern, O. (1955). *Theory of games and economic behavior* (2nd rev. ed.). Princeton University Press.
3. Poundstone, W. (1992). *Prisoner's Dilemma: John von Neumann, game theory, and the puzzle of the bomb*. Anchor Books.
4. Poundstone, W. (1992). *Prisoner's Dilemma: John von Neumann, game theory, and the puzzle of the bomb*. Anchor Books.
5. Poundstone, W. (1992). *Prisoner's Dilemma: John von Neumann, game theory, and the puzzle of the bomb*. Anchor Books.
   See also: Siracusa, J. M. (2008). *Nuclear weapons: A very short introduction*. Oxford University Press.
6. Poundstone, W. (1992). *Prisoner's Dilemma: John von Neumann, game theory, and the puzzle of the bomb*. Anchor Books.
7. Princeton. (n.d.) *AC105 Nash John Forbes 1950*. Seeley G. Mudd Manuscript Library. Dokumen. https://dokumen.tips/documents/ac105-nash-john-forbes-1950.html
8. Nash, J. (1951). Non-Cooperative Games. *Annals of Mathematics*, *54*(2), 286-295. https://doi.org/10.2307/1969529
9. The Nobel Prise Organization (1994, October 11). *The Sveriges Riksbank Prize in Economic Sciences in memory of Alfred Nobel 1994*. https://www.nobelprize.org/prizes/economic-sciences/1994/press-release/
10. Selten, R. (1975). Reexamination of the perfectness concept for equilibrium points in extensive games. *International Journal of Game Theory*, *4*(1), 25–55. https://doi.org/10.1007/bf01766400
11. The Nobel Prise Organization (1994, October 11). *The Sveriges Riksbank Prize in Economic Sciences in memory of Alfred Nobel 1994*. https://www.nobelprize.org/prizes/economic-sciences/1994/press-release/

12. Harsanyi, J. C. (1982). Games with Incomplete Information Played by "Bayesian" Players, I–III Part I. The Basic Model. In *Papers in Game Theory*, (pp. 115–138). Springer, Dordrecht. https://doi.org/10.1007/978-94-017-2527-9_6

8. Practice Makes Permanent - How To Never Forget What You Learn

1. Roediger, H. L., & Karpicke, J. D. (2006). Test-Enhanced Learning: Taking memory tests improves long-term retention. *Psychological Science*, *17*(3), 249–255. https://doi.org/10.1111/j.1467-9280.2006.01693.x

9. Making A Playbook - Why Humans And Robots Are Different

1. Jasonzweig. (2017, October 2). *What Harry Markowitz meant*. Jason Zweig. https://jasonzweig.com/what-harry-markowitz-meant/
2. Cope, J. M. (2011). Entrepreneurial learning from failure: An interpretative phenomenological analysis. *Journal of Business Venturing*, *26*(6), 604–623. https://doi.org/10.1016/j.jbusvent.2010.06.002
3. Stocks, E., Lishner, D. A., & Decker, S. K. (2008). Altruism or psychological escape: Why does empathy promote prosocial behavior? *European Journal of Social Psychology*, *39*(5), 649–665. https://doi.org/10.1002/ejsp.561
4. Pham, L.B., & Taylor, S. E. (1999). From Thought to Action: Effects of Process-Versus Outcome-Based Mental Simulations on Performance. *Personality and Social Psychology Bulletin*, *25*(2), 250–260. https://doi.org/10.1177/0146167299025002010
5. Ostrom, E., Burger, J., Field, C. B., Norgaard, R. B., & Policansky, D. (1999). Revisiting the Commons: Local Lessons, Global Challenges. *Science*, *284*(5412), 278–282. https://doi.org/10.1126/science.284.5412.278

DISCLAIMER

The information contained in this book and its components, is meant to serve as a comprehensive collection of strategies that the author of this book has done research about. Summaries, strategies, tips and tricks are only recommendations by the author, and reading this book will not guarantee that one's results will exactly mirror the author's results.

The author of this book has made all reasonable efforts to provide current and accurate information for the readers of this book. The author and their associates will not be held liable for any unintentional errors or omissions that may be found, and for damages arising from the use or misuse of the information presented in this book.

Readers should exercise their own judgment and discretion in interpreting and applying the information to their specific circumstances. This book is not intended to replace professional advice (especially medical advice, diagnosis, or treatment). Readers are encouraged to seek appropriate professional guidance for their individual needs.

The material in the book may include information by third parties. Third party materials comprise of opinions expressed by their owners. As such, the author of this book does not assume responsibility or liability for any third party material or opinions.

The publication of third party material does not constitute the author's guarantee of any information, products, services, or opinions contained within third party material. Use of third party material does not guarantee that your results will mirror our results. Publication of such third party material is simply a recommendation and expression of the author's own opinion of that material.

Whether because of the progression of the Internet, or the unforeseen changes in company policy and editorial submission guidelines, what is stated as fact at the time of this writing may become outdated or inapplicable later.

Wisdom University is committed to respecting copyright laws and intellectual property rights. We have taken reasonable measures to ensure that all quotes, diagrams, figures, images, tables, and other information used in this publication are either created by us, obtained with permission, or fall under fair use guidelines. However, if any copyright infringement has inadvertently occurred, please notify us promptly at wisdom-university@mail.net, providing sufficient details to identify the specific material in question. We will take immediate action to rectify the situation, which may include obtaining necessary permissions, making corrections, or removing the material in subsequent editions or reprints.

This book is copyright ©2023 by Wisdom University with all rights reserved. It is illegal to redistribute, copy, or create derivative works from this book whole or in parts. No parts of this report may be reproduced or retransmitted in any forms whatsoever without the written expressed and signed permission from the publisher.

Made in the USA
Las Vegas, NV
13 May 2024